The Art of Proving:
Algorithmic and Visual Approaches in Discrete Mathematics

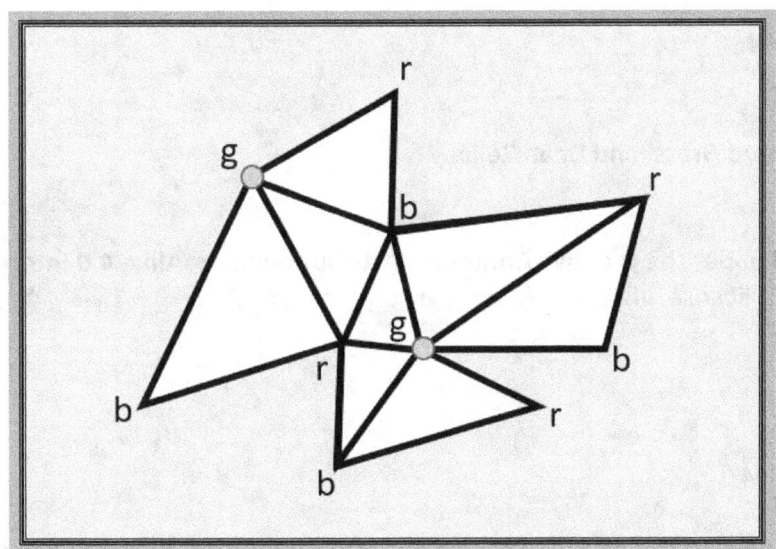

Richard Grassl   **Dean Zeller**

Richard Grassl
Emeritus Professor of Mathematical Sciences
University of Northern Colorado
Greeley, CO 80639

Dean Zeller
Lecturer, Mathematical Sciences
University of Northern Colorado
Greeley, CO 80639

© 2020-2021 by Richard Grassl and Dean Zeller

This work is licensed under the Creative Commons Attribution-ShareAlike 4.0 International License. To view a copy of this license, visit http://creativecommons.org/licenses/by-sa/4.0/

1st Edition

ISBN: 979-8732298994

A current electronic version can be found for free at http://www.openmathbooks.org/proving

Prepared for publication by Oscar Levin for Open Math Books

Cover artwork by artist and mathematician KAJ. Untitled: Critters from Design Motifs of Ancient Mexico, by Jorge Encisco, 1947; Present Tense Conjugation, in Kinyarwanda, of the verb "to have". And, of course, the picture proof.

Contents

Forewords *5*
- Foreword by Dean Zeller 5
- Foreword by Richard Grassl 7
- Foreword by Sydney Small 10
- Acknowledgements 11

Introduction – Don't Jump to Conclusions! *13*
- Example 1 – Polynomial Always Returns Prime Numbers 13
- Example 2 – Triangular Numbers 16
- Example 3 – Line Segment Lengths 16
- Example 4 – Odd Numbers of the Form $m^2 + n^2$ 17
- Example 5 – Is the number 33333..331 always prime? 18

Chapter 1 – Telescoping Sums and Products *19*

Chapter 2 – Gems *25*
- Gem 1 – Consecutive Composite Numbers 26
- Gem 2 – Up-Down Permutations 28
- Gem 3 – Making Triangles 31
- Gem 4 – No Squares Allowed 37
- Gem 5 – How Odd? 37
- Gem 6 – The Art Gallery Theorem 38
- Gem 7 – Chromatic Triangles 40
- Gem 8 – How Many (little) Squares? 42
- Gem 9 – Greatest Postage Value Unable to Make 44
- Gem 10 – Maximum Number of Regions 46
- Gem 11 – Which is Bigger? 49
- Gem 12 – Counting Rectangles 51
- Gem 13 – Counting More Rectangles 53
- More Gems For You 54

Chapter 3 – Combinatorial Proofs *57*

Chapter 4 – Visual Approach *71*
- (De)Compose Yourself 72
- Poster Instructions – Matching 74

Chapter 5 – Computer Analysis *81*
- **Proof Using Code** 81
- **Proof Using Existing Software** 84
- **Problems** 86
- **Solutions** 87

Chapter 6 – Induction *103*

Appendix 1 – Poster Cutouts *131*

Appendix 2 – Finale *135*

About the Authors *137*

Forewords

Foreword by Dean Zeller

So I'm in my office, minding my own business, and suddenly my office-neighbor Richard Grassl started asking me questions about drawing graphics for a discrete mathematics textbook he was working on. Nothing complicated – just circles, lines, and squares placed to demonstrate mathematical problems visually. However, being a mathematics professor for 50+ years, Richard didn't know what was possible with drawing graphics or how to do it.

Then I come along. While my degrees are in mathematics and computer science, my personal interests are much more diverse. I write movie scripts for the fun of it, knowing Michael Bay will never be interested in directing them. I design lessons in wilderness survival, martial arts, and culinary, despite having little formal training in any of them. And yes, graphic design is an interest. For a while, it was a profession, having taught courses in graphic design, web design, and interior design at the Art Institute of Jacksonville. Looking at Grassl's request for some circles, squares, and lines, I realized that it was calling to me. I had joined in previous mathematics-related endeavors at UNC, including multiple mathematics contests, science fairs, and presentations. I saw this as an opportunity.

The book you hold in your hand (or reading on a screen) is a collection of ideas by the mathematical expertise of Dr. Richard Grassl. He presents new ways of looking at problems, at various learning styles, including the use of technology or looking at problems algorithmically. This is 50 years' worth of mathematical ideas.

As for me, the book needed an algorithmic point of view, in addition to drawing pretty pictures and organizing the overall document. While there are approximately 48 million Discrete Mathematics textbooks, give or take 47.999 million, all of them have the same content – discrete mathematics. This book is not intended to be the mainstay of a discrete mathematics course. We simply wanted to create a discussion on how to solve problems in different ways. Most textbooks focus on content and technique, which is fine. This book discusses alternate methods for tackling problems, whether traditionally, with graphic diagrams, or with a computer program. Using these methods, students may be able to solve a greater range of problems than before. This book is an exploration into this goal.

Discrete mathematics consists of lots of witty, clever, and easy-to-ask questions, but with sometimes bafflingly difficult solutions. Wherever possible, Python programs were written to help illustrate the point, as well as to discuss how programming can be used as a method of mathematical proof.

It is a common educational observation that not all students learn in the same way. With different brains, there are different learning styles. Many educational psychologists have studied this phenomenon. One of the most accepted theories are those of Howard Gardner's Multiple Intelligences. The idea is that teachers must be ready to deal with different learning styles among students. This is not to say that a lesson should be written many times for different learning styles, but teachers should design lessons with the idea that students with different learning styles will be using the material. It is with this pedagogical philosophy in mind that we offer to the mathematical community a very useful resource. Textbook? No, it's a workbook. Nah, that sounds too elementary. Just what should we call this work?

Foreword by Richard Grassl

One day, I woke up thinking about my 57 years of teaching and dreamt of organizing and compiling some of the interesting and fun mathematics that I have been privileged to research and teach. A lot of mathematics is simply very inspiring fun, with a little bit of challenge added. Reading about mathematics, trying to solve interesting mathematical problems (for example: how would you factor $x^5 + x + 1$?), and engaging in thought-provoking challenges that mathematics presents is a welcome relief from non-understandable, often contradictory, and irrational behavior that politics, religion, or even sports present.

This initial attempt captures just a slice of the types of problems that are encountered in undergraduate mathematics. I mainly focus on just one particular branch of mathematics – discrete and combinatorial mathematics – with a tiny flavor of what K-12 students competing in mathematics contests often experience.

This document is neither a textbook nor a set of *Cliff Notes*, but rather, a compilation of our ideas and thoughts. Let's call it a BOOK!

This book was written for a variety of audiences and purposes. It provides a different look to traditional mathematics for undergraduates, for preservice teachers, in-service teachers, for faculty that teach discrete mathematics, computer science, and also special mathematics appreciation courses that satisfy general education requirements. A collaboration between a mathematician and an instructor of computer science helped reveal how technology-supported investigations could open doors to proving, and to understanding.

A beginning instructor of Mathematics Induction will appreciate the completely worked solutions to a large variety of n-problems. Secondary teachers along with professors can display simpler and contrasting proving techniques using Telescoping Sums/Products. A Combinatorial Proof looks at one idea two different ways. Often the gorilla glue that cements steps together is a <u>Computer Analysis</u>. Combining the <u>Visual Approach</u> along with data collection, enhanced by a computer program enables one to plow forward toward a satisfying conclusion.

Attacking problems with these 5-6 hammers can broaden one's grip and extend mental capacity and willingness to dig deeper. "How many diagonals in a convex n-gon?" A start with induction might transition over to a more direct and even easier geometric perspective. Our many different proofs of $\binom{m+n}{2} - \binom{m}{2} - \binom{n}{2} = mn$ will appeal to those who think more visually, to those that appreciate tedium and direct algebra, and to those appealing to real life (try m men and n women). And then there are always those who will try induction. Although most of the problems and solutions presented have a discrete flavor, our message is independent of topics chosen. Keep an open mind, draw a picture, approach the problem from different directions, gather data using computer analysis, and then glue it all together.

With that in mind, here is an example of five different proofs that verify a very common inequality.

THEOREM: Let $x > 0$ be any real number. Then $x + \frac{1}{x} \geq 2$.

1. A standard calculus proof is quick. Let $f(x) = x + \frac{1}{x}$, $f'(x) = 0$ when $x = 1$. The second derivative test confirms that f has a minimum of 2 at x=1.

2. A graph of $y = x + \frac{1}{x}$ pretty clearly shows a minimum at $x = 1$.

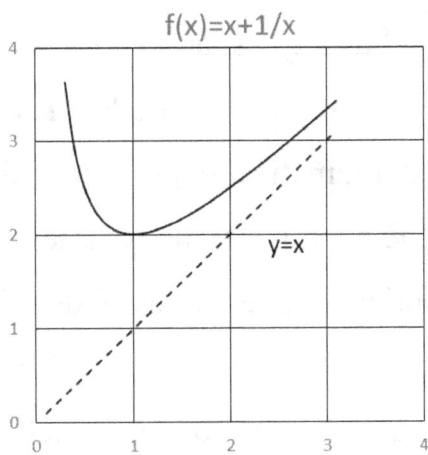

3. The Arithmetic – Geometric mean inequality, where $a_i > 0$,

$$(a_1 + a_2 + a_3 + \ldots + a_n) \div n \geq \sqrt{a_1 a_2 a_3 \ldots a_n}$$

works, but it is pretty heavy duty. You get $\frac{x + \frac{1}{x}}{2} \geq \sqrt{x \frac{1}{x}} = 1$ and then $x + \frac{1}{x} \geq 2$.

4. It is clear that $\left(\sqrt{x} - \frac{1}{\sqrt{x}}\right)^2 \geq 0$. Then $x - 2 + \frac{1}{x} \geq 0$ finishes it.

5. Picture proof with no words: You should go ahead and interpret this picture.

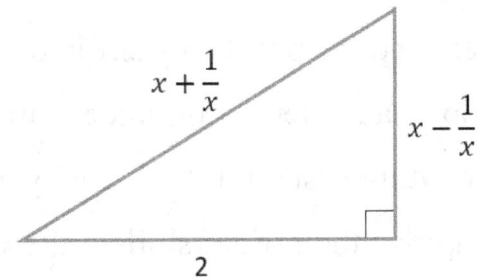

Foreword by Sydney Small

Working with Dr. Grassl as a typist and editor for this book has been a much greater experience than I had initially bargained for. I was tasked with the entry of written proofs and complex mathematical expressions, formatting the text and visuals, and editing parts of the document (many times.) All of this combined made for quite the educational experience, one that extended far beyond simply editing a book.

As an undergraduate mathematics student myself, I am able to appreciate Dr. Grassl's way of breaking down proofs in order to help readers understand his thought process. For me, having the ability to effectively read and write proofs is what really opened the door to higher mathematics. However, this was also one of the most difficult things to accomplish – in fact, I am still working on improving my proof skills to this day. This is something I know is true for many students alike. The learning curve for proving ideas in mathematics is anything but black and white; understanding how one theorem works does not guarantee the understanding of others. As such, the best thing a student can do to broaden their skills is to read as many proofs as possible, then make their own attempts. Editing this book allowed me to do just that.

As I was going through each part to make sure everything read as intended, I realized that I had to understand the proofs myself in order to truly edit the work. Now, while it is not your job as a reader to eliminate comma splices and wrestle with Microsoft Word's formatting system, I think this type of thorough read-through can help anyone improve their proof skills. Likewise, I have found having multiple ways to view a concept can be incredibly helpful, especially when learning something as complex as this. This may come in the form of verbal description, algebraic representation, or even a picture. For some of the proofs in this book, Dean Zeller demonstrated how they could even be proved through computer programming, an unnerving method to most mathematicians, including myself. However, this just goes to show how multi-faceted discrete mathematics truly is. While this book only scratches the surface of the vast world of mathematical proofs, there is so much contained in these pages that can be appreciated by proof novices and experts alike. I know I certainly have.

Acknowledgements

Dedicated to my mother, who brought out my love of teaching and the developing mind. And in the memory of my father, who motivated my sheer love of logic, research, and a strong sense of ethics. Richard and Joan, this book is for you. (Dean Zeller)

Dedicated to the nuns at Mora Central HS in Watsonville, CA, who started me on this journey, and to Prof. Abraham P. Hillman, my PhD advisor, coauthor, and mentor for nearly half-a-century. (Richard Grassl)

Introduction – Don't Jump to Conclusions!

Why do we need proofs? Well, many times we can be confronted with overwhelming data that indicates something to be true, but upon further investigation turns out to be false.

As we proceed examining various types of proofs in discrete (and sometimes in mathematics in general) be aware that not all proofs provide a satisfying feeling. Consider this question: if a and b are irrational numbers, can a^b ever be rational? Consider the irrational number $\sqrt{2}$ and look at the two disjoint cases: Either $\sqrt{2}^{\sqrt{2}}$ is rational or irrational. If rational, we are done. If irrational, then look at

$$\left(\sqrt{2}^{\sqrt{2}}\right)^{\sqrt{2}} = \sqrt{2}^2 = 2.$$

Here are several classic examples of statements that look to be true, but in fact are false, as counter examples will show.

Example 1 – Polynomial Always Returns Prime Numbers

Consider the polynomial $p(n) = n^2 - n + 41$. Starting at 0, start computing values.

n	0	1	2	3	4	5	6	7	8	9
p(n)	41	41	43	47	53	61	71	83	97	113

Initial Conjecture

Students usually make two observations here. First, successive differences are 0, 2, 4, 6, 8, 10, …, a fact that is easy to prove.

$$p(n+1) - p(n) = n^2 + n + 41 - (n^2 - n + 41) = 2n$$

Second, it appears that each number produced is a <u>prime</u> number. Make a few more. Try p(10), p(11), … In fact, p(n) is prime for n = 0, 1, 2, …, 40, but fails when n = 41 since p(41) = 41² − 41 + 41 = 41². One counterexample is sufficient to disprove a claim. So here we have a situation where an observation is true for 41 cases but is in fact false in general. There are other such polynomials that yield even more "convincing" data.

Programming

Finding the counterexample took calculating the formula 42 times. By hand, this can be quite tedious. However, the code below will repeatedly calculate the formula for numbers and indicate whether the number is prime. While programming may not always be appropriate for verifying a proof, it is quite effective at finding counterexamples.

Finding Prime Numbers

```
def p(n):
    result = n*n - n + 41
    return result

def test_P_on_range(start, end):
    "Calculates p and determines if it is prime"
    print("Calculaing",p.__doc__,"from",start,"to",end)
    print("   n     p(n)    status")
    for i in range(start, end+1):
        calc = p(i)
        if isPrime(calc):
            message = "prime"
        else:
            message = "composite"
        print("%4d  %6d  %s" % (i, calc, message))

def isPrime(n):
    "Brute-force algorithm to check for prime numbers"
    for i in range(2,n//2):
        if n % i == 0:
            return False
    return True

test_P_on_range(0,45)
```

The output of the above code confirms the results that 0 through 40 are correct, but 41 and others present counterexamples.

```
Calculating p(n) = n^2 - n + 41 from 0 to 45

n    p(n)    status                n    p(n)    status
 0     41    prime                 23     547   prime
 1     41    prime                 24     593   prime
 2     43    prime                 25     641   prime
 3     47    prime                 26     691   prime
 4     53    prime                 27     743   prime
 5     61    prime                 28     797   prime
 6     71    prime                 29     853   prime
 7     83    prime                 30     911   prime
 8     97    prime                 31     971   prime
 9    113    prime                 32    1033   prime
10    131    prime                 33    1097   prime
11    151    prime                 34    1163   prime
12    173    prime                 35    1231   prime
13    197    prime                 36    1301   prime
14    223    prime                 37    1373   prime
15    251    prime                 38    1447   prime
16    281    prime                 39    1523   prime
17    313    prime                 40    1601   prime
18    347    prime                 41    1681   composite
19    383    prime                 42    1763   composite
20    421    prime                 43    1847   prime
21    461    prime                 44    1933   prime
22    503    prime                 45    2021   composite
```

The program took some time to write, but it is likely comparable to calculating the formula 45 times, and certainly more enjoyable. This problem-solving method can also be adopted to many other proofs, as you will experience reading upcoming chapters.

Example 2 – Triangular Numbers

Determine pairs (m,n), with m and n positive integers, such that
$1! \cdot 3! \cdot 5! \cdot 7! \cdot \ldots \cdot (2m-1)! = n!$.

When m=1, n=1 so our first pair is (1,1). When m=2, $1! \cdot 3! = 3!$, so n=3 and (2,3) is a second pair. For m=3, $1! \cdot 3! \cdot 5! = 6!$ So we get (3,6). For m=4, the factors of $1! \cdot 3! \cdot 5! \cdot 7!$ can be rearranged into 10! Giving us (4,10). We have produced 1, 3, 6, 10, the first four triangular numbers leading us to conjecture that 15 is the next number n. But try to make 15! out of $1! \cdot 3! \cdot 5! \cdot 7! \cdot 9!$. You cannot make an 11 or a 13! (This last ! is a note of surprise, not a factorial.) In fact these are the only values that work, as nice as they are. Displaying these for easier reading:

$$1! = 1!$$
$$1! \cdot 3! = 3!$$
$$1! \cdot 3! \cdot 5! = 6!$$
$$1! \cdot 3! \cdot 5! \cdot 7! = 10!$$

"Factorials appear like tiny bursts of joy in a landscape of numbers."

Example 3 – Line Segment Lengths

Determine all the different lengths of a line segment on an n×n pin geoboard, for n=2, 3, 4, … . For n=2, you get just two different lengths, 1 and $\sqrt{2}$. For n=3, you get five: 1, $\sqrt{2}$, 2, $\sqrt{5}$, and $2\sqrt{2}$. There are nine for n=4, and fourteen for n=5. This is a nice sequence: 2, 5, 9, 14, and a reasonable guess for n=6 would be 20, since the differences are 3, 4, 5, and 6. But in fact, there are only 19 because of the Pythagorean triple $3^2 + 4^2 = 5^2$; you get 5 counted twice.

The nice pattern that we were hoping for would look like this:

2 5 9 14 20 27 35 44 54 65 77 90 104 119

where the differences are fairly predictable. But the 6 by 6 grid of dots has the number 5 duplicated due to the presence of a 3 – 4 – 5 triangle in the 5 by 5 grid. This persists until the 6 – 8 – 10 Pythagorean triple shows up in the 9 by 9 grid so that the 11 by 11 grid has an extra 10. Next, the 5 – 12 – 13 triple shows up in the 13 by 13 grid so 13 is repeated in the 14 by 14 grid.

So the above sequence really looks like:

2 5 9 14 19 26 34 43 53 64 75 88 102 116

Example 4 – Odd Numbers of the Form $m^2 + n^2$

As you start listing such numbers it appears like you are generating terms of the arithmetic sequence 1, 5, 9, 13, … with common difference $d = 4$. But look what happens.

$0^2 + 1^2 = 1$ $\qquad$ $2^2 + 5^2 = 29$

$1^2 + 2^2 = 5$ $\qquad$ $ = 33$

$0^2 + 3^2 = 9$ $\qquad$ $1^2 + 6^2 = 37$

$2^2 + 3^2 = 13$ $\qquad$ $4^2 + 5^2 = 41$

$1^2 + 4^2 = 17$ $\qquad$ $3^2 + 6^2 = 45$

$ = 21$ $\qquad$ $0^2 + 7^2 = 49$

$0^2 + 5^2 = 25$ $\qquad$ $2^2 + 7^2 = 53$

$\qquad\qquad\qquad\qquad = 57$

The initial pattern falls apart; you cannot make certain numbers like 21, 33, 57,… by summing two terms of the sequence of squares 1, 4, 9, 16, 25, 36, 49,…

The reader might want to explore this idea and to try to determine the nature of the numbers that you cannot make.

Example 5 – Is the number 33333..331 always prime?

The question remains, is the number 333…331 always prime? Before resorting to writing a computer program, let's start and collect some data.

$$\begin{aligned}
31 &\quad \text{is prime} \\
331 &\quad \text{is prime} \\
3331 &\quad \text{is prime} \\
33331 &\quad \text{is prime} \\
333331 &\quad \text{is prime} \\
3333331 &\quad \text{is prime} \\
33333331 &\quad \text{is prime}
\end{aligned}$$

Looks like we have a winner! But hold on. 333333331 is NOT a prime. In fact, it factors as $17 \cdot 19607843$, with a little help from technology (although Euler factored such numbers long ago by hand). You cannot trust such patterns.

You need proof! Polya once said "An intuition is not proof."

Chapter 1 – Telescoping Sums and Products

The telescoping sum, or product, technique is nearly magical in its simplicity. You just need to be able to convert a product into a difference such as:

$$\frac{1}{3 \cdot 4} = \frac{1}{3} - \frac{1}{4}$$

$$F_{n+2}F_{n-1} = F_{n+1}^2 - F_n^2$$

$$\ln\left(\frac{x}{y}\right) = \ln(x) - \ln(y),$$

$$3 \cdot 3! = 4! - 3!.$$

Sometimes something as simple as $F_n = F_{n+1} - F_{n-1}$, where F_n is the nth Fibonacci number, does the trick. Now, let's start:

A. $\frac{1}{1 \cdot 2} + \frac{1}{2 \cdot 3} + \cdots + \frac{1}{n(n+1)} = \frac{n}{n+1}$

Decompose the LHS into:

$$\left(\frac{1}{1} - \frac{1}{2}\right) + \left(\frac{1}{2} - \frac{1}{3}\right) + \left(\frac{1}{3} - \frac{1}{4}\right) + \cdots + \left(\frac{1}{n} - \frac{1}{n+1}\right) = 1 - \frac{1}{n+1} = \frac{n}{n+1}$$

Look at the cancellations! Only two terms remain.

B. $\ln\left(\frac{1}{2}\right) + \ln\left(\frac{2}{3}\right) + \ln\left(\frac{3}{4}\right) + \cdots + \ln\left(\frac{n}{n+1}\right)$

$= (\ln(1) - \ln(2)) + (\ln(2) - \ln(3)) + (\ln(3) - \ln(4)) + \cdots + (\ln(n) - \ln(n+1))$

$= \ln(1) - \ln(n+1)$

$= \ln\left(\frac{1}{n+1}\right).$

We could also accomplish this using telescoping products.

$$\text{LHS} = \ln\left(\frac{1}{2} \cdot \frac{2}{3} \cdot \frac{3}{4} \cdots \frac{n}{n+1}\right) = \ln\left(\frac{1}{n+1}\right)$$

C. $F_1 + F_3 + F_5 + \cdots + F_n$
$= (F_2 - F_0) + (F_4 - F_2) + (F_6 - F_4) + \cdots + (F_{n+1} - F_{n-1}) = F_{n+1} - F_0$
$= F_{n+1}$

(For n odd)

D. $(1 \cdot 1!) + (2 \cdot 2!) + (3 \cdot 3!) + \cdots + (n \cdot n!)$
$= (2! - 1!) + (3! - 2!) + (4! - 3!) + (5! - 4!) + \cdots + ((n+1)! - n!)$
$= (n+1)! - 1!$

This one depends on the algebraic step $n \cdot n! = (n+1)! - n!$

E. $F_0 F_3 + F_1 F_4 + F_2 F_5 + F_3 F_6 + \cdots + F_{n-1} F_{n+2}$
$= (F_2^2 - F_1^2) + (F_3^2 - F_2^2) + (F_4^2 - F_3^2) + (F_5^2 - F_4^2) + \cdots$
$\quad + (F_{n+1}^2 - F_n^2)$
$= F_{n+1}^2 - 1$

It is easy to first prove that $F_{n+1}^2 - F_n^2 = F_{n+2} F_{n-1}$

F. $F_1^2 + F_2^2 + F_3^2 + \cdots + F_n^2$
$= F_1(F_2 - F_0) + F_2(F_3 - F_1) + F_3(F_4 - F_2) + \cdots + F_n(F_{n+1} - F_{n-1})$
$= -F_1 F_0 + F_n(F_{n+1})$
$= F_n F_{n+1}$

Here we use the easy to prove fact that

$$F_n^2 = F_n(F_{n+1} - F_{n-1})$$

G. $\dfrac{1}{1\cdot 3} + \dfrac{1}{3\cdot 5} + \dfrac{1}{5\cdot 7} + \cdots + \dfrac{1}{(2n-1)(2n+1)}$

$= \dfrac{1}{2}\left[\dfrac{1}{1} - \dfrac{1}{3} + \dfrac{1}{3} - \dfrac{1}{5} + \dfrac{1}{5} - \dfrac{1}{7} + \cdots + \dfrac{1}{2n-1} - \dfrac{1}{2n+1}\right]$

$= \dfrac{1}{2}\left[1 - \dfrac{1}{2n+1}\right] = \dfrac{n}{2n+1}$

This partial fraction decomposition will also work nicely on

$$\dfrac{1}{1\cdot 4} + \dfrac{1}{4\cdot 7} + \dfrac{1}{7\cdot 10} + \cdots + \dfrac{1}{(3n-2)(3n+1)}$$

where the factors are 3 apart.

H. $\left(1 - \dfrac{2}{3}\right)\left(1 - \dfrac{2}{5}\right)\left(1 - \dfrac{2}{7}\right)\cdots\left(1 - \dfrac{2}{2n+1}\right) = \dfrac{1}{3}\cdot\dfrac{3}{5}\cdot\dfrac{5}{7}\cdot\dfrac{7}{9}\cdots\dfrac{2n-1}{2n+1} = \dfrac{1}{2n+1}$

I. $(1+x)(1+x^2)(1+x^4)(1+x^8)\cdots = \dfrac{1-x^2}{1-x}\cdot\dfrac{1-x^4}{1-x^2}\cdot\dfrac{1-x^8}{1-x^4}\cdot\dfrac{1-x^{16}}{1-x^8}\cdots = \dfrac{1}{1-x}$

With proper interpretation, this identity can be used to show that every positive integer n can be uniquely represented in binary form.

J. $(1 + x + x^2 + \cdots + x^9)(1 + x^{10} + x^{20} + \cdots + x^{90})(1 + x^{100} + x^{200} + \cdots + x^{900})\ldots$

$= \dfrac{1-x^{10}}{1-x}\cdot\dfrac{1-x^{100}}{1-x^{10}}\cdot\dfrac{1-x^{1000}}{1-x^{100}}\cdots = \dfrac{1}{1-x}$

Base 10 representation is unique!

K. Let's try to show $2[F_0 + F_3 + F_6 + \cdots + F_{3n}] = F_{3n+2} - 1$

LHS $= F_0 + F_1 + F_2 \quad + F_4 + F_5 \quad + F_7 + F_8 + \cdots + F_{3n-2} + F_{3n-1} +$
$(F_0 + F_3 + F_6 + \cdots + F_{3n})$ using the recursion for F_n,
$= F_0 + F_1 + F_2 + F_3 \ldots + F_{3n}$, filling in the "holes,"
$= (F_3 - F_2) + (F_4 - F_3) + (F_5 - F_4) + \cdots + (F_{3n+2} - F_{3n+1})$
$= -F_2 + F_{3n+2}$
$= F_{3n+2} - 1$

L. $1^2 - 2^2 + 3^2 - 4^2 + 5^2 - 6^2 + \cdots + (n-1)^2 - n^2$
$= (1-2)(1+2) + (3-4)(3+4) + (5-6)(5+6) + \cdots$
$\quad + (n-1-n)(n-1+n)$
$= (-1)[1 + 2 + 3 + 4 + \cdots + (n-1) + n]$
$= -1 \cdot \frac{n(n+1)}{2}$ for n even.

M. $\frac{1}{2!} + \frac{2}{3!} + \frac{3}{4!} + \cdots + \frac{n}{(n+1)!}$
$= \left(\frac{1}{1!} - \frac{1}{2!}\right) + \left(\frac{1}{2!} - \frac{1}{3!}\right) + \left(\frac{1}{3!} - \frac{1}{4!}\right) + \cdots + \left(\frac{1}{n!} - \frac{1}{(n+1)!}\right)$
$= 1 - \frac{1}{(n+1)!}$

This decomposition uses the fact that $\frac{n}{(n+1)!} = \frac{1}{n!} - \frac{1}{(n+1)!}$

N. Here we show $1^2 + 2^2 + 3^2 + \cdots + n^2 = \frac{n(n+1)(2n+1)}{6}$

Expand $\sum_{x=1}^{n} x^3 - (x-1)^3$ two ways and then equate them:

$\sum x^3 - (x-1)^3$
$= (1^3 - 0^3) + (2^3 - 1^3) + (3^3 - 2^3) + \cdots + (n^3 - (n-1)^3)$
$= n^3.$

Next, $\sum x^3 - (x-1)^3 = 3\sum x^2 - 3\sum x + \sum 1 = 3\sum x^2 - \frac{3(n)(n+1)}{2} + n$

From $n^3 = 3\sum x^2 - \frac{3n(n+1)}{2} + n$, you get $\sum x^2 = \frac{n(n+1)(2n+1)}{6}$.

This technique is general, since $\sum x^k$ can be found once you know $\sum x^{k-1}$.

O. Solve the recursion $a_n = a_{n-1} + 3$

$a_n - a_{n-1} = 3$
$a_{n-1} - a_{n-2} = 3$
$a_{n-2} - a_{n-3} = 3$
$\vdots$
$a_2 - a_1 = 3$

Now add both columns to get: $a_n - a_1 = (n-1) \cdot 3$

and finally $a_n = a_1 + (n-1) \cdot 3$.

P. This one is a little harder. Find a nice formula for $\frac{1}{3} + \frac{2}{21} + \frac{3}{91} + \cdots + \frac{n}{n^4 + n^2 + 1}$.

We can create the following difference and then factor:

$$n^4 + n^2 + 1 = n^4 + 2n^2 + 1 - n^2 = (n^2 + 1)^2 - n^2$$

$$= (n^2 - n + 1)(n^2 + n + 1).$$

Then, $\frac{n}{n^4+n^2+1} = \frac{1}{2}\left[\frac{1}{n^2-n+1} - \frac{1}{n^2+n+1}\right]$

Now we can telescope:

$$\frac{1}{3} + \frac{2}{21} + \frac{3}{91} + \cdots + \frac{n}{n^4+n^2+1}$$

$$= \frac{1}{2}\left[\left(1 - \frac{1}{3}\right) + \left(\frac{1}{3} - \frac{1}{7}\right) + \left(\frac{1}{7} - \frac{1}{13}\right) + \cdots + \left(\frac{1}{n^2-n+1} - \frac{1}{n^2+n+1}\right)\right]$$

$$= \frac{1}{2}\left[1 - \frac{1}{n^2+n+1}\right] = \frac{n(n+1)}{2(n^2+n+1)}.$$

Q. Here are three you can try:

$$\left(1 - \frac{1}{4}\right)\left(1 - \frac{1}{9}\right)\left(1 - \frac{1}{16}\right) \cdots \left(1 - \frac{1}{n^2}\right) = ?$$

$$(x+1)(x^2+1)(x^4+1)(x^8+1) \cdots (x^{1024}+1) = ?$$

$$\left(1 + \frac{1}{2}\right)\left(1 + \frac{1}{3}\right)\left(1 + \frac{1}{4}\right) \cdots \left(1 + \frac{1}{n}\right) = ?$$

Chapter 2 – Gems

When Richard was in the 8th grade, the nun teaching mathematics asked this question in class: "How many consecutive composite integers can you have?" Inspired by the clever-exhilarating-and very simple solution (see Gem 1), he knew, at that moment that he wanted to be a mathematician.

Later, the question "which line has more points?"

and its solution

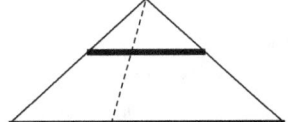

helped seal the deal. This was a very early introduction to the idea of a one-to-one correspondence (every line segment from the apex to the base of the triangle connects exactly two points).

A GEM is a mathematical problem with some unique magical charisma that holds the power to charm an eager mind into a state of commitment. Simple statements like "every natural number greater than 11 can be written as a sum of two composite numbers" can be a lure to students at a very young age.

Gem 1 – Consecutive Composite Numbers

How much would you be willing to bet against me if I said I could produce 999 consecutive composite (positive) numbers? It is easy to get 3 in a row; 8/9/10 and 14/15/16 work correctly. It is also not too difficult to get 5 in a row: 24/25/26/27/28. Can you get exactly 4 or 6 in a row? Here are two 7-in-a-row: 90/91/92/93/94/95/96 and 120/121/122/123/124/125/126. Now let's see what Python can produce. Here is the code:

```python
# Gem -- Consecutive Composite Numbers

import math

def isPrime(n):
    "Brute-force algorithm to check for prime numbers"
    for i in range(2,n//2+1):
        if n % i == 0:
            return False
    return True


def consecCompositeVerbose(start, end):
    cons = []
    for i in range(start, end+1):
        if isPrime(i):
            print()
            print("Prime: ",str(i)+":   ",len(cons),"previous composites   ",cons)
            cons=[]
        else:
            cons.append(i)

def consecCompositeSummary(start, end):
    comps = []
    maxComps = 1
    compsStr = ""
    print("# in a row     consecutive composites          ending prime")
    for i in range(7, end+1):
        if isPrime(i):
            if len(comps)>maxComps:
                maxComps = len(comps)
                if len(comps)<=5:
                    compsStr = ""
                    for i in range(len(comps)-1):
```

```
                            compsStr += str(comps[i])+", "
                        compsStr += str(comps[i+1])
                    print(" %4d         %-36s    %5d" % (len(comps),
compsStr ,i))
                else:
                    print(" %4d         %-36s    %5d" % (len(comps),
str(comps[0])+", "+str(comps[1])+", ... , "+str(comps[-1]),i))
            comps=[]
        else:
            comps.append(i)


#consecCompositeVerbose(2,542)
consecCompositeSummary(7,1000000)
```

And here is the output:

```
# in a row        consecutive composites              ending prime
    3         8, 9, 10                                     1
    5         24, 25, 26, 27, 28                           3
    7         90, 91, ... , 96                            97
   13         114, 115, ... , 126                        127
   17         524, 525, ... , 540                        541
   19         888, 889, ... , 906                        907
   21         1130, 1131, ... , 1150                    1151
   33         1328, 1329, ... , 1360                    1361
   35         9552, 9553, ... , 9586                    9587
   43         15684, 15685, ... , 15726                15727
   51         19610, 19611, ... , 19660                19661
   71         31398, 31399, ... , 31468                31469
```

The results are not unique in the sense that there are other sequences of consecutive composite numbers of a particular length. By hand, the above would be very difficult to locate.

Observation 1: You cannot get exactly any particular even number in a row.

Observation 2: Not all lengths exists; for example, there is no sequence of length exactly 11. We grade this computer analysis with an A-.

Not bad, but take a look at this. Here are 999 in a row:

$$1000!+2, 1000!+3, 1000!+4, \ldots, 1000!+1000$$

They are clearly consecutive and composite: 2 divides the first term, 3 divides the second term, and so forth until 1000 divides the last term. With this technique, you can produce 9999 or more consecutive composite numbers.

Gem 2 – Up-Down Permutations

A permutation of the integers 1, 2, 3, 4, ..., n which first rises and then alternately falls and rises again is called an up-down (U-D) sequence or an up-down permutation. For example, 264513 is one of the U-D sequences for n = 6. Let E_n denote the number of up-down sequences for 1, 2, 3, ..., n. These first few values for E_n are not too hard to compute by hand:

$E_1 = 1$

$E_2 = 1$ 12

$E_3 = 2$ 132   231

$E_4 = 5$ 1324   1423   2314   2413   3412

$E_5 = 16$ 13254   14253   14352   15342   15243   23154
            24153   24351   25143   25341   34152   34251
            35142   35241   45132   45231

The data table looks like this:

n	1	2	3	4	5	6	7	8	9	10
E_n	1	1	2	5	16	61	272	1385	7936	50521

The first five columns were calculated by hand, but the last five required the program below. This program could not be run on n more than 10, as there were too many permutations for the computer memory. Work for the future could include an algorithm that does not have to create all of the permutations at the start.

```
# Gem -- UpDown Permutations

from itertools import permutations

def isUpDown(perm):
    "Determines whether a permutation is an UpDown"
    upDown = True   #assume true, prove otherwise
    current = perm[0]
    for i in range(1,len(perm)):
        if i%2==1:  # up
            if perm[i]<current:
                upDown = False
                break
            current = perm[i]
        else:  # down
            if perm[i]>current:
                upDown = False
                break
            current = perm[i]
    return upDown


def numUpDownsOnN(n):
    nums = []
    for i in range(n):
        nums.append(i+1)
    perms = list(permutations(nums))

    numUpDowns = 0
    for perm in perms:
        if isUpDown(perm):
            numUpDowns += 1

    print("For n =",n,"there are",numUpDowns,"UpDown permutations.")


def main():
    for i in range(1,15):
        numUpDownsOnN(i)

main()
```

The output is below. At n=11, the program had a memory error, due to the size of 11! permutations. For further work, we can investigate algorithms that generate the permutations on the fly, rather than generating them in a list.

```
For n = 1 there are 1 UpDown permutations.
For n = 2 there are 1 UpDown permutations.
For n = 3 there are 2 UpDown permutations.
For n = 4 there are 5 UpDown permutations.
For n = 5 there are 16 UpDown permutations.
For n = 6 there are 61 UpDown permutations.
For n = 7 there are 272 UpDown permutations.
For n = 8 there are 1385 UpDown permutations.
For n = 9 there are 7936 UpDown permutations.
For n = 10 there are 50521 UpDown permutations.
```

One punch line! Those of you that teach calculus may have already recognized these numbers. These numbers, E_n, are called Euler numbers, and appear in the Maclaurin series expansion for secx + tanx. First let $E_0 = 1$.

$$\sec x + \tan x = 1 + x + \frac{x^2}{2!} + 2\frac{x^3}{3!} + 5\frac{x^4}{4!} + 16\frac{x^5}{5!} + 61\frac{x^6}{6!} + 272\frac{x^7}{7!} + \ldots$$

One more calculus connection. Suppose p(x) is a polynomial of degree 5, such that its derivative has four real, distinct roots. In how many ways can we prescribe the orders in which the 4 distinct extremes can be arranged? Each of these graphs below start with a (relative) maximum, followed by alternating minima and maxima. These charts were created in Microsoft Excel, using the Line Graph tool. To flip the 1-2-3-4's for typical graphs, negative numbers were used.

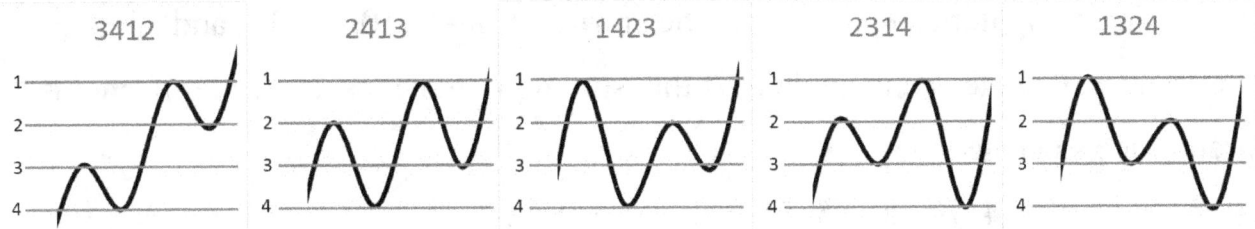

Above each graph we listed one of the corresponding $E_4 = 5$ up-down permutations. This illustration generalizes perfectly. The reader might want to draw the $E_5 = 16$ risings and fallings of the maxima and minima for a polynomial of degree 6.

Gem 3 – Making Triangles

Let T(n) be the number of ways of selecting three distinct numbers from the set {1, 2, 3, ... ,n} so that they are the lengths of the sides of a triangle. As an example T(5) = 3; the only possibilities are 2-3-4, 2-4-5, and 3-4-5. How about a formula for T(n)?

As stated, one could imagine a student in grades 4, 5 or 6 trying this problem. That and the complicated and intriguing results that grow out of the data earn the privilege of membership in the GEM group.

First collect data and make a table by hand:

n	3	4	5	6	7	8	9	10	11	12	13
T(n)	0	1	3	7	13	22	34	50	70	95	125

For example, the 7 triangles for n = 6 are:

234   245   256   345   346   356   456

The computer analysis fills in the entries for n= 9, 10, 11, 12, and 13. One observation that we can make here is that summing two consecutive terms yields a pretty nice sequence:

1, 4, 10, 20, 35, 56, 84, 120, 165,...

(Thanks to the program for providing the extended data). These are the tetrahedral numbers from the 4th diagonal of the Pascal triangle. In notation, T(n + 1) + T(n) = $\binom{n}{3}$. While this is a nice result, determining a general formula is not so easy so let's call it done!.

Now, suppose you add to the statement of this problem that for every choice of 3 from the set {1, 2, 3, ..., n} you must include n, now how many triangles can you make? By hand a partial table is not too difficult to construct:

n	4	5	6	7	8	9	10	11	12	13
T(n)	1	2	4	6	9	12	16	20	25	30

The last 5 entries were generated by computer analysis.

The six for n = 7 are:

267   357 367   457 467 567.

Notice the 1-2-2-1 grouping. The nine for n = 8 are:

278   368 378   458 468 478   568 578   678

with a 1-2-3-2-1 grouping. The respective groups here start with 2, 3, 4, 5 and 6. The reader might want to try n=9 showing a 1-2-3-3-2-1 grouping. The computer analysis now extends the data table from 9 to 13.

```
# Gem -- Making Triangles

def makeTriangles(n):
    print("Making triangles for n =",n)
    if n>=10:
        print(" (Using hex A thru F for lengths 10 to 15)")
    if n>=16:
        print(" (Triangles will not look correct for lengths")
        print("greater than 15, but results are still counted)")
    triangles = []
    for i in range(1, n+1):
        for j in range(1, n+1):
            for k in range(1, n+1):
                if i != j and i != k and j != k:
                    if i+j>k and i+k>j and j+k>i:
                        newTriangle = [i,j,k]
                        newTriangle.sort()
                        if newTriangle not in triangles \
                            and newTriangle[2]==n:
                                triangles.append(newTriangle)
    for t in triangles:
        for side in t:
            print(hex(side)[2:].upper(), end="")
        print(end=" ")
    print()
    print("Number of triangles for n =",n,"->",len(triangles))
    print()


def main():
    for i in range(4,14):
        makeTriangles(i)

main()
```

The output of the program can be compared to the table above for accuracy. For readability purposes, hexadecimal digits were used for side lengths of 10 or more.

```
Making triangles for n = 4
234
Number of triangles for n = 4 -> 1

Making triangles for n = 5
245 345
Number of triangles for n = 5 -> 2

Making triangles for n = 6
256 346 356 456
```

```
Number of triangles for n = 6 -> 4

Making triangles for n = 7
267 357 367 457 467 567
Number of triangles for n = 7 -> 6

Making triangles for n = 8
278 368 378 458 468 478 568 578 678
Number of triangles for n = 8 -> 9

Making triangles for n = 9
289 379 389 469 479 489 569 579 589 679 689 789
Number of triangles for n = 9 -> 12

Making triangles for n = 10
(Using hex A thru F for lengths 10 to 15)
29A 38A 39A 47A 48A 49A 56A 57A 58A 59A 67A 68A 69A 78A 79A 89A
Number of triangles for n = 10 -> 16

Making triangles for n = 11
(Using hex A thru F for lengths 10 to 15)
2AB 39B 3AB 48B 49B 4AB 57B 58B 59B 5AB 67B 68B 69B 6AB 78B 79B
7AB 89B 8AB 9AB
Number of triangles for n = 11 -> 20

Making triangles for n = 12
(Using hex A thru F for lengths 10 to 15)
2BC 3AC 3BC 49C 4AC 4BC 58C 59C 5AC 5BC 67C 68C 69C 6AC 6BC 78C
79C 7AC 7BC 89C 8AC 8BC 9AC 9BC ABC
Number of triangles for n = 12 -> 25

Making triangles for n = 13
(Using hex A thru F for lengths 10 to 15)
2CD 3BD 3CD 4AD 4BD 4CD 59D 5AD 5BD 5CD 68D 69D 6AD 6BD 6CD 78D
79D 7AD 7BD 7CD 89D 8AD 8BD 8CD 9AD 9BD 9CD ABD ACD BCD
Number of triangles for n = 13 -> 30
```

The grouping pattern can now be extended as follows:

n = 4	1	1
n = 5	1 1	2
n = 6	1 2 1	4
n = 7	1 2 2 1	6
n = 8	1 2 3 2 1	9
n = 9	1 2 3 3 2 1	12
n = 10	1 2 3 4 3 2 1	16
n = 11	1 2 3 4 4 3 2 1	20

The identity

$$1 + 2 + 3 + 4 + \ldots + (n-1) + n + (n-1) + \ldots + 3 + 2 + 1 = n^2$$

with its corresponding picture with n=4 is shown next. Each diagonal has 1, then 2, then 3, and so forth, and the eventual result looks like a (squished) square.

$$T(10) = 1 + 2 + 3 + 4 + 3 + 2 + 1 = 16 = \left(\frac{10-2}{2}\right)^2$$
$$\uparrow$$
$$n$$

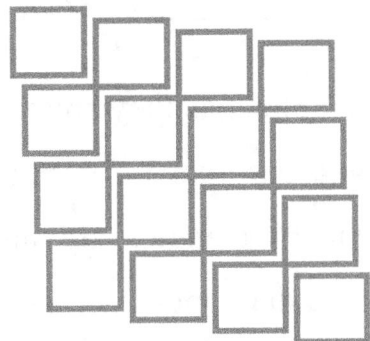

This picture can help determine a general formula for n even.

For n even, $T(n) = \left(\frac{n-2}{2}\right)^2$

When n is odd, use $1 + 2 + 3 + \ldots + n = \frac{n(n+1)}{2}$ or $2(1 + 2 + \ldots + n) = n(n+1)$.

$$T(11) = 1 + 2 + 3 + 4 + 4 + 3 + 2 + 1 = 2 \cdot 10 = 20$$

For n odd, $T(n) = \left(\frac{n-3}{2}\right)\left(\frac{n-3}{2} + 1\right)$

Continuing with n=11, $T(11) = \left(\frac{8}{2}\right)\left(\frac{8}{2} + 1\right) = 4 \cdot 5$

Numbers of this form, n(n+1), are called OBLONG numbers. On the diagram that follows, you should add the total number of dots above an angle.

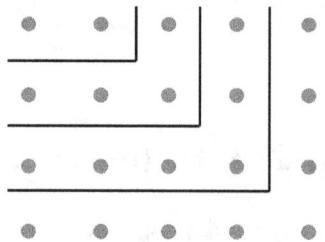

Look at the T(n) when n is odd on this "extended" data table:

n	5	7	9	11	13	15	17	19
T(n)	2	6	12	20	30	42	56	72

These numbers are called oblong as they are one off from being square:

$$1 \cdot 2,\ 2 \cdot 3,\ 3 \cdot 4,\ 4 \cdot 5,\ 5 \cdot 6,\ 6 \cdot 7,\ 7 \cdot 8,\ 8 \cdot 9$$

They each make an n by n+1 rectangle. Fifth and 6th graders would like this.

This gem continues to yield diamond-like (or at least ore-like) results.

Examine the sums of two consecutive terms in the "extended" data table for all n:

$1 + 2 = \quad 3 = t_2$ (t_2 is a triangular number)

$2 + 4 = \quad 6 = t_3$

$4 + 6 = \quad 10 = t_4$

$6 + 9 = \quad 15 = t_5$

$9 + 12 = \quad 21 = t_6$

$12 + 16 = \quad 28 = t_7$

$16 + 20 = 36 = t_8$

And now we have a recursion for T(n) in terms of the triangular numbers:

$$T(n) + T(n+1) = t_{n-2} = \frac{(n-2)(n-1)}{2}.$$

Gem 4 – No Squares Allowed

Problem: Show that the product of four consecutive natural numbers cannot be a perfect square.

Data collection is a must! Look at n(n+1)(n+2)(n+3) for a bunch of n's.

n	1	2	3	4	5	6
Product	24	120	360	840	1680	

If you know your squares each of these seem to be 1 less than a perfect square, and so cannot be a square. Now, look at what happens when you add 1 in general:

$$n(n+1)(n+2)(n+3) + 1 = n^4 + 6n^3 + 11n^2 + 6n + 1 = (n^2 + 3n + 1)^2.$$

It looks like our conjecture holds in general.

Gem 5 – How Odd?

Problem: When is the binomial coefficient $\binom{n}{k}$ odd?

On the Pascal triangle several early rows consist of all odd numbers. Perhaps we should start there.

n = 3   1 3 3 1

n = 7   1 7 21 35 35 21 7 1

n = 15   1 15 105 455 1365 3003 5005 6435 6435 5005 3003 1365 455 105 15 1

These row numbers are interesting in that they are each one less than a power of 2; this might be enough to prompt looking at base 2 representation. When you think of even-odd one often thinks of parity – powers of 2 – base 2 – binary.

$$1 = 1_2$$
$$3 = 11_2$$
$$7 = 111_2$$
$$15 = 1111_2$$

Now look at each $\binom{n}{k}$ for n = 7 in binary form.

$$\binom{111}{000} \binom{111}{001} \binom{111}{010} \binom{111}{011} \binom{111}{100} \binom{111}{101} \binom{111}{110} \binom{111}{111}$$

Each of these 8 integers share one property – the top number, 111, dominates the bottom number, in the sense that there is never a 0 directly above a 1. This answers the original question:

$\binom{n}{k}$ is odd whenever n in binary form dominates k in binary form.

Try it! Is $\binom{37}{11}$ even or odd? Look at $\binom{100101}{001011}$.

The proof in general is hard. We are just trying to give you some insight derived from the data. One can only imagine what a computer program would look like.

Gem 6 – The Art Gallery Theorem

This is a famous problem from the *Heart of Mathematics* book. An art gallery floor plan is typically a region where the walls form a polygonal path. The corners where the walls meet are called vertices. Where should security cameras be placed so that every point in the gallery can be viewed by at least one camera? All cameras swivel perfectly. Here is the interesting part: we want as few cameras as possible!

The simplest floor plan is a triangle; any one of the three vertices can serve as a camera site. Now, let's start with the statement of the theorem and try to understand it. If the polygonal closed path has v vertices then at most v/3 vertices will suffice for sites of the security camera. If v/3 is not an integer, round down.

So here is the plan:

Step 1 – Triangulate the shape. Each triangle will have three of the vertices of the polygonal path as vertices.

Step 2 – Color the v vertices of the path with red, blue, or green so that the 3 vertices of any triangle each have a different color.

Now we have $R + B + G = v$, where R, B, and G stand for the numbers of red, blue, or green vertices. Remember that each of the v vertices were colored exactly once. At least one of R, G, B must be less than or equal to $v/3$. Now we have it. Choose any point inside the gallery. It is in some triangle. Now choose either all red, or all blue, or all green vertices for placement of the cameras, whichever total ≤ 3. Below is a typical floor plan, triangulated with arbitrary coloring. Place the cameras at the two green vertices.

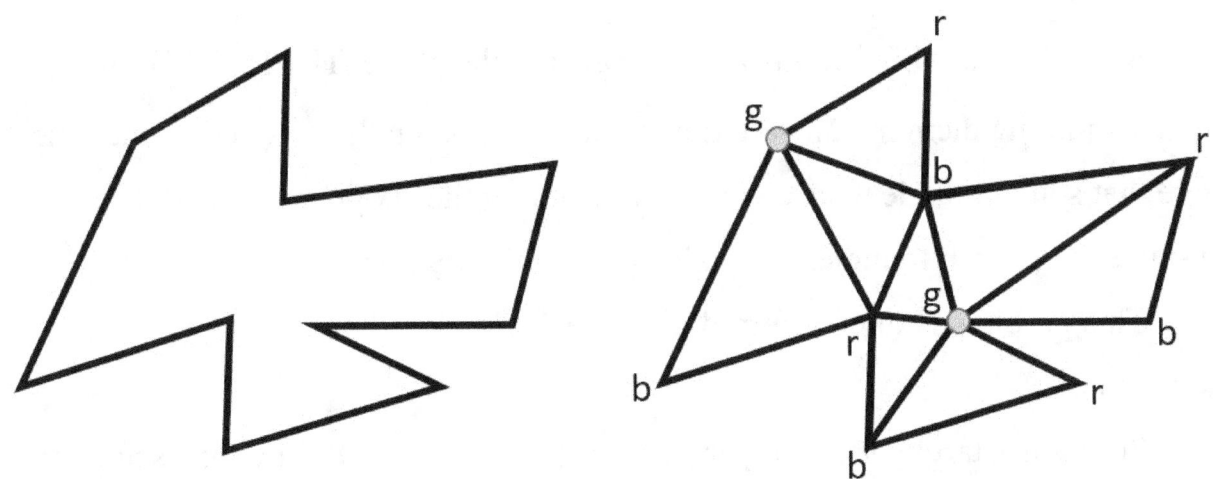

Here are a few art gallery floor plans that you can work on. What is the minimum number of cameras or guards needed so that the gallery is secure from theft?

Now let's turn the question around. Draw a floor plan of an art gallery with 10 sides that requires exactly two guards to view the entire gallery.

Now draw a floor plan for an art gallery with 9 sides that requires exactly 3 guards to watch the entire gallery. Also show where they should be placed.

Gem 7 – Chromatic Triangles

Six points, no three collinear, are drawn in the plane. The $15 = \binom{6}{2}$ line segments joining them in pairs are drawn, and then painted, some red, some blue. Prove that some triangle whose vertices are among the six points has all its sides the same color; such triangles are called <u>chromatic triangles.</u>

This problem is often referred to as the 3 friends – 3 strangers problem in many texts.

Solution: Start by selecting any one of the 6 points and draw line segments to each of the other 5 points.

At least three of these lines must be the same color, say red; and it certainly doesn't matter which 3 lines (you could rearrange them). If any one of the 3 dotted lines are red we are done. So make them all blue; but then we have a blue-blue-blue chromatic triangle.

Notice that if we only have 5 points this result is false. Color the 5 perimeter segments red and the 5 interior lines blue; no chromatic triangle is produced.

There is an interesting extension of this problem. You are about to continue to experience a glimpse into Ramsey Theory.

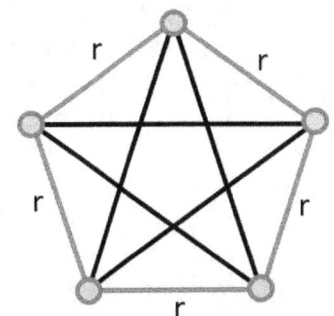

Suppose you have 17 points in the plane and again no three points are collinear. Draw the $\binom{17}{2}$ line segments connecting each pair as before, and color them with 3 colors, red, blue, and green. Prove that no matter how you color them you will be forced to produce a chromatic triangle.

Start as before and select one of the 17 points and draw the 16 line segments to each of the other 16 points. Color them red, blue, or green as at random as you can. At least 6 of these 16 segments must be colored the same color, say red. Otherwise, at best you would have 5 red, 5 blue, and 5 green. What color must the 16th segment be colored? Answer: red or blue or green.

Now you have $15 = \binom{6}{2}$ segments to be colored blue or green (remember to not paint these red as you are trying to not make a chromatic triangle). A few of the 16 dotted segments are shown below. But now we are done as this 17 point and 3 color problem is now narrowed down to the previous 6 point, 2 color problem.

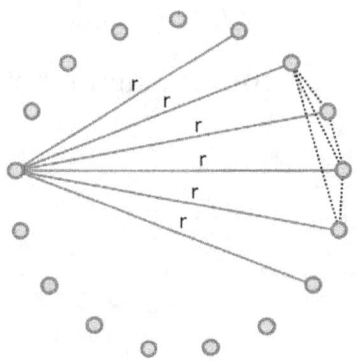

41

Gem 8 – How Many (little) Squares?

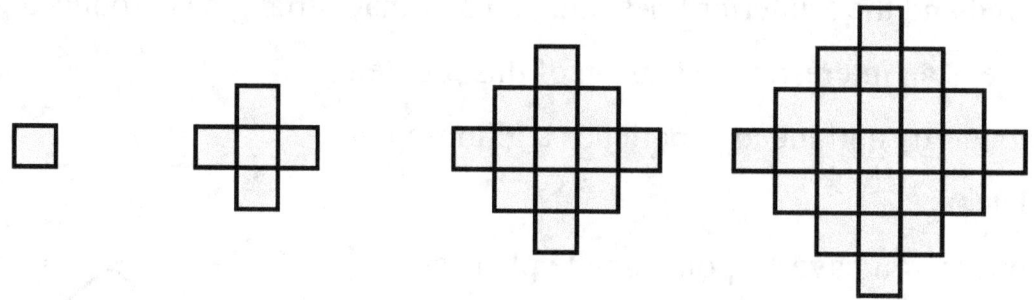

If you draw the next few diagrams you will produce the sequence 1, 5, 13, 25, 41, 61, ... giving the total number of squares per diagram. The number of squares on each row for the first five diagrams are:

									Total
				1					1
			1	3	1				5
		1	3	5	3	1			13
	1	3	5	7	5	3	1		25
1	3	5	7	9	7	5	3	1	41

Now, let's make use of one of our "picture proofs." The sum $1+3+5+7=4^2$ comes from the figure below,

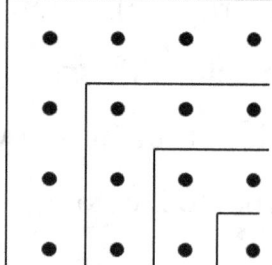

42

leading us to notice $1 + 3 = 4$, $1 + 3 + 5 = 9$, $1 + 3 + 5 + 7 = 16$ and finally we have the following data from the "split" table:

$$1 = 1$$
$$5 = 1 + 4$$
$$13 = 4 + 9$$
$$25 = 9 + 16$$
$$41 = 16 + 25$$
$$61 = 25 + 36$$

First observation: the nth diagram, which has $2n-1$ squares in the middle row, has $n^2 + (n-1)^2$ total squares.

Second observation: You could also embed each diagram into a larger square, and subtract the corners:

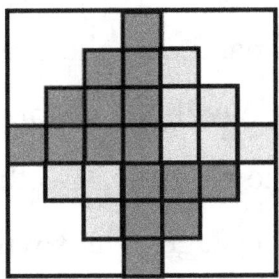

In general, $(\text{odd})^2 - 4(\text{a triangular number}) = (2n-1)^2 - 4(\frac{(n-1)n}{2}) = n^2 + (n-1)^2$.

Third observation – A stunning turn: Roll the triangular chunk up to the left and fill in the gap; repeat for the bottom smaller chunk, rolling to the right, creating two squares as before.

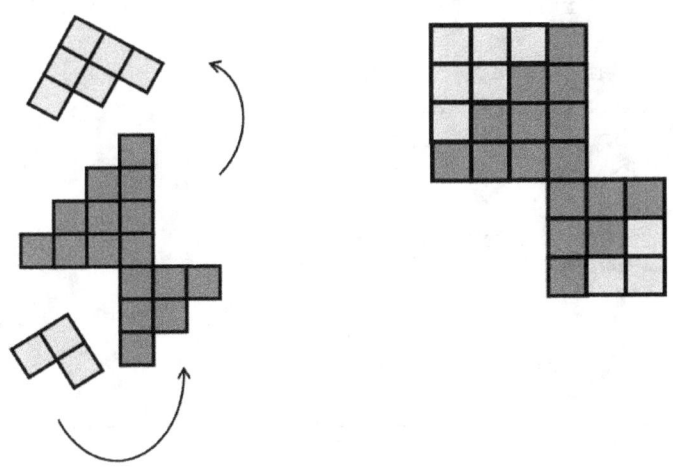

Gem 9 – Greatest Postage Value Unable to Make

What is the largest amount of postage that you cannot make using 5-cent and 9-cent stamps? Do the same with 15¢ and 29¢ stamps. This program can be applied to most two stamp-values. (Don't try 4¢ and 6¢ stamps.)

Programming this solution was actually quite fun, in that it was not nearly as complicated as originally thought. Simply put, the program creates 100 Boolean variables in an array, all set to false. That size can be increased for other stamp values. The 0th Boolean is set to true, to indicate it is a possible configuration of stamps. That's when the inductive magic takes place. Starting at the beginning, look through each Boolean value. If the value is false, do nothing, as the index represents a value that cannot be made with the stamps. If it is true, then it represents a value that can be made with the stamps. As such, the value plus each of the possible stamp values is also possible. Do that from index 0 up through the end of the array, and all possible postage values will be represented. This algorithm could easily be extended to include more than just two possible stamp values.

```
def stampValues(a, b, arraySize):
    bools = []
    for i in range(arraySize+a+b):
        bools.append(False)

    bools[0] = True

    for i in range(arraySize):
        if bools[i]==True:
            bools[i+a]=True
            bools[i+b]=True

    for i in range(arraySize):
        print("%3d %5s" % (i,bools[i]))

stampValues(5,9, 35)
stampValues(18,29, 500)
```

The resulting output is below. Note that the last false value is 31, which is the answer to the 5-cent and 9-cent stamp problem. After that, all values will be true. In fact from (5-1) (9-1) = 4 · 8 = 32 on all postage values can be made. Actually if you get 5 true values in a row, you could simply start adding 5. For instance if 34 is true, so is 39. After this point all of them will be true.

0	True		20	True
1	False		21	False
2	False		22	False
3	False		23	True
4	False		24	True
5	True		25	True
6	False		26	False
7	False		27	True
8	False		28	True
9	True		29	True
10	True		30	True
11	False		31	False
12	False		32	True
13	False		33	True
14	True		34	True
15	True		35	True
16	False		36	True
17	False		37	True
18	True		38	True
19	True		39	True

The program was run again with values of 15 and 29 cents. An excerpt of the output is below. Note that 391 is the last false value, and all greater values will be true. Also note that (15 - 1) (29 - 1) = 14 · 28 = 392. In fact, it is true that for a¢ and b¢ stamps, all values from (a - 1)(b - 1) can be made, as long as a and b are relatively prime. With small stamp values, this is a wonderful problem for younger students.

370	True		382	True
371	True		383	True
372	True		384	True
373	True		385	True
374	True		386	True
375	True		387	True
376	False		388	True
377	True		389	True
378	True		390	True
379	True		391	False
380	True		392	True
381	True		393	True

Gem 10 – Maximum Number of Regions

Problem: determine the maximum number of regions into which a plane is separated by n lines. To guarantee the maximum number, no two lines can be parallel, and no three lines concurrent.

Let P(n) denote the maximum number of regions created by n lines. If n=0, there is just one region, so P(0) = 1. If n=1, there are two regions, so P(1) = 2. Drawing pictures will produce the following data:

n	0	1	2	3	4	5	6
P(n)	1	2	4	7	11	16	22

P(1) = 2   P(2) = 4   P(3) = 7   P(4) = 11

Here is what n = 3 looks like:

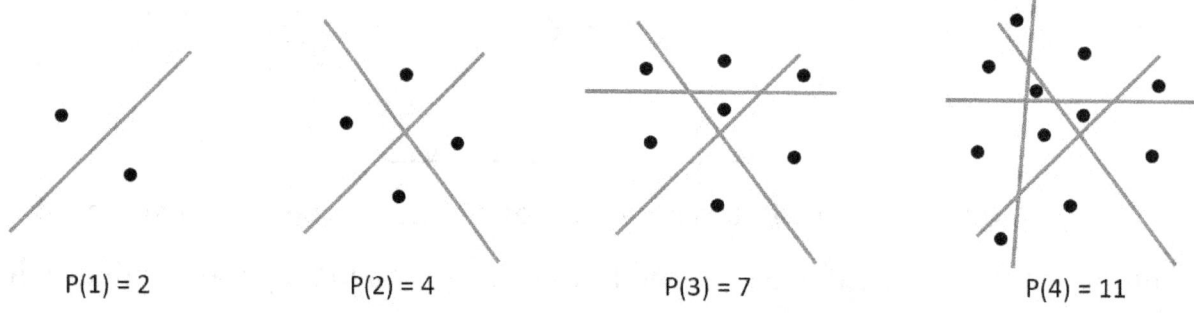

When the (dotted) 4th line is drawn in, it splits 4 regions, creating 4 new regions. Equivalently, P(4) = P(3) + 4, and the same for n = 5.

The resultant recursion P(n) = P(n-1) + n can be solved using telescoping sums to yield $P(n) = \frac{n(n+1)}{2} + 1$. You could also spot this from the data table above. Each entry is 1 more than a triangular number.

But this is not the end of the story! The plane is two-dimensional. What is the analogous problem in one dimension and in three dimensions? How many lines (segments), L(n), do n points divide a line into? And what is the maximal number of regions, S(n), that n planes divide 3-space into?

The first question is easy to visualize; draw a line and slap on a bunch of dots.

Here, n = 5 points divide the line into 6 segments.

This following table is loaded with data involving L(n), P(n), and S(n).

n	L(n)	P(n)	S(n)
0	1	1	1
1	2	2	2
2	3	4	4
3	4	7	8
4	5	11	15
5	6	16	$\boxed{26}$
6	7	$\boxed{22}$	

It is pretty easy to see that $L(n) = n + 1$. And we already had the formula $P(n) = \frac{n(n+1)}{2} + 1$. But here is the problem. Data collection for S(n) is not easy, but if you notice certain patterns in the table a formula is revealed:

$$L(2) + P(2) = P(3) \quad 3 + 4 = 7$$
$$L(3) + P(3) = P(4) \quad 4 + 7 = 11$$
$$L(4) + P(4) = P(5) \quad 5 + 11 = 16$$
$$L(5) + P(5) = P(6) \quad 6 + 16 = \boxed{22}$$

and we could jump to

$$P(2) + S(2) = S(3) \quad 4 + 4 = 8$$
$$P(3) + S(3) = S(4) \quad 7 + 8 = 15$$

and then to $11 + 15 = 26$, the entry below the 15 in the last column. The resulting recursion $S(n) = S(n-1) + P(n-1)$ can be telescoped to achieve $S(n) = \binom{n+1}{3} + n + 1$. Of course, this is still a conjecture and a proof is required; you should try this.

An additional unexpected pattern!

There are enough binomial coefficients floating around to motivate reconfiguring our three formulas as follows:

$$L(n) = n + 1 = \qquad\qquad \binom{n}{1} + \binom{n}{0}$$
$$P(n) = \frac{n(n+1)}{2} + 1 = \binom{n+1}{2} + 1 = \binom{n}{2} + \binom{n}{1} + \binom{n}{0}$$
$$S(n) = \binom{n+1}{3} + n + 1 = \qquad \binom{n}{3} + \binom{n}{2} + \binom{n}{1} + \binom{n}{0}$$

A remarkable pattern!

A good detective could easily conjecture a formula for the maximum number of regions that 3-space divides 4-space. But actually counting a few beginning cases might be hard.

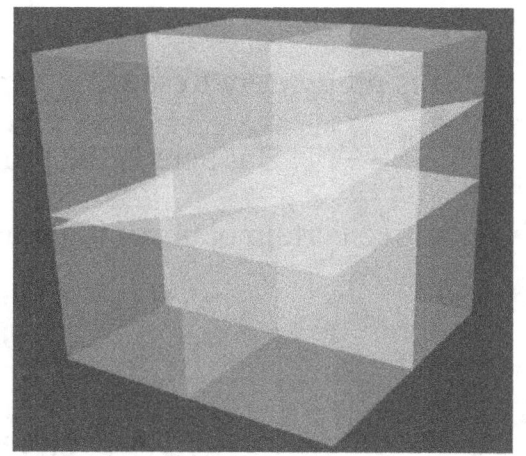

Gem 11 – Which is Bigger?

$$1 + \frac{1}{2} + \frac{1}{4} + \frac{1}{8} + \ldots \quad \text{or} \quad 1 + \frac{1}{2} + \frac{1}{3} + \frac{1}{4} + \frac{1}{5} + \ldots$$

The first one is a geometric sum and we know how to evaluate it. Let

$$S = 1 + \frac{1}{2} + \frac{1}{4} + \frac{1}{8} + \ldots = 1 + \frac{1}{2}(1 + \frac{1}{2} + \frac{1}{4} + \ldots) = 1 + \frac{1}{2}S. \text{ Now solve for S:}$$

$$S = 2$$

For the second one, notice that

$$\frac{1}{3} + \frac{1}{4} > \frac{1}{4} + \frac{1}{4} = \frac{1}{2}$$

$$\frac{1}{5} + \frac{1}{6} + \frac{1}{7} + \frac{1}{8} > \frac{4}{8} = \frac{1}{2}$$

$$\frac{1}{9} + \frac{1}{10} + \frac{1}{11} + \frac{1}{12} + \frac{1}{13} + \frac{1}{14} + \frac{1}{15} + \frac{1}{16} > \frac{8}{16} = \frac{1}{2}$$

and so on. The next 16 terms add up to more than another ½.; the next 32 terms add up to more than another ½…. The program below demonstrates the two sums. The first one, geometricSum, converges to 2.0 after 54 iterations, and all future iterations will also be 2.0. In contrast, the program calculates how many iterations are required to reach a harmonicSum of 8.0, 9.0, and 10.0. Each solution took significantly longer to execute.

CONCLUSION: The HARMONIC series $1 + \frac{1}{2} + \frac{1}{3} + \frac{1}{4} + \frac{1}{5} + \ldots$ diverges to ∞. You can make the sum bigger than any number by simply adding up enough ½ 's.

The program below produces the output discussed.

```
# Gem -- Which series is larger, geometric or harmonic sum

import math

def geometricSum(numTerms):
    total = 0
    for i in range(numTerms):
        total += 1/2**i
    return total

def harmonicSum(numTerms):
    total = 0
    for i in range(1,numTerms):
        total += 1/i
    return total

def main():
    print("Testing Geometric sum to reach 2")
    for i in range(1000000):
        x = geometricSum(i)
        print(i, x)
        if x==2.0:
            break

    print()
    print("Testing Harmonic sum to reach certain values.")
    for i in range(1000000):
        x = harmonicSum(i)
        if x>8:
            print(i, x)
            break
    for i in range(1000000):
        x = harmonicSum(i)
        if x>9:
            print(i, x)
            break
    for i in range(1000000):
        x = harmonicSum(i)
        if x>10:
            print(i, x)
            break

main()
```

The output is as follows:

```
Testing Geometric sum to reach 2
0 0
1 1.0
2 1.5
3 1.75
4 1.875
5 1.9375
6 1.96875
7 1.984375
8 1.9921875
9 1.99609375
10 1.998046875
11 1.9990234375
12 1.99951171875
13 1.999755859375
14 1.9998779296875
15 1.99993896484375
16 1.999969482421875
17 1.9999847412109375
18 1.9999923706054688
19 1.9999961853027344
20 1.9999980926513672
21 1.9999990463256836
22 1.9999995231628418
23 1.999999761581421
24 1.9999998807907104
25 1.9999999403953552
26 1.9999999701976776
27 1.9999999850988388
28 1.9999999925494194
29 1.9999999962747097
30 1.9999999981373549
31 1.9999999990686774
32 1.9999999995343387
33 1.9999999997671694
34 1.9999999998835847
35 1.9999999999417923
36 1.9999999999708962
37 1.999999999985448
38 1.999999999992724
39 1.999999999996362
40 1.999999999998181
41 1.9999999999990905
42 1.9999999999995453
43 1.9999999999997726
44 1.9999999999998863
45 1.9999999999999432
46 1.9999999999999716
47 1.9999999999999858
48 1.999999999999993
49 1.9999999999999964
50 1.9999999999999982
51 1.9999999999999991
52 1.9999999999999996
53 1.9999999999999998
54 2.0

Testing Harmonic sum to reach
certain values.
1675 8.000485571995782
4551 9.000208062931115
12368 10.000043008275778
```

Interpretation: you need to add all the way up to 1/12368 just to get to 10. Imagine how long it would just to get to 11, or 37.

Gem 12 – Counting Rectangles

Based on student and teacher responses the next pair are tied for BEST GEM. How many rectangles of all sizes are there in a subdivided rectangular m×n grid? Here squares count as rectangles and a 1×2 shape is counted as different from a 2×1 shape, for example. Let's try a direct approach and count them by size.

The following analysis of a 3×4 rectangle can be easily extended to the general case,

Size	#	Size	#	Size	#
1×1	12	2×1	8	3×1	4
1×2	9	2×2	6	3×2	3
1×3	6	2×3	4	3×3	2
1×4	3	2×4	2	3×4	1

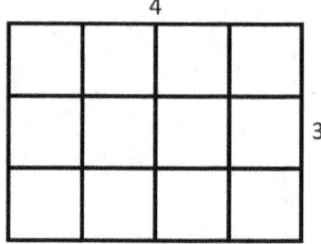

Now add these, being alert for any surprising patterns:

$$(1 + 2 + 3 + 4) + 2(1 + 2 + 3 + 4) + 3(1 + 2 + 3 + 4) =$$

$$(1 + 2 + 3)(1 + 2 + 3 + 4) =$$

$$\frac{3 \cdot 4}{2} \cdot \frac{4 \cdot 5}{2} = 6 \cdot 10 = 60.$$

Once the hard work is complete, remember what Polya said – "If you arrive at a nice answer, perhaps there is a nice solution lurking close by". Here it is! The above answer can be written as $\binom{4}{2} \cdot \binom{5}{2}$. Now interpret these numbers as follows: Choose any two of the four horizontal lines along with any two of the five vertical lines.

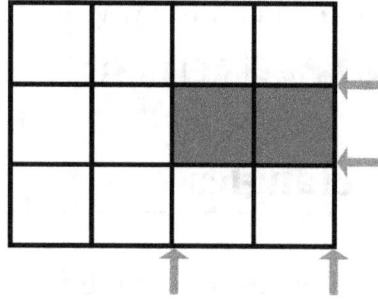

That choice will uniquely determine a rectangle; conversely any choice of a rectangle directly corresponds to exactly two line choices. This slick solution was made possible by first collecting and then organizing the data. For an m×n rectangle, that same idea shows that there are $\binom{m+1}{2}\binom{n+1}{2}$ rectangles of all sizes.

Gem 13 – Counting More Rectangles

This last problem is a close relative. How many rectangles are there in this 4-tableau shape? The n-tableau shape will have a similar look. For example, the 5-tableau shape will have a row of five 1-by-1 squares on top of this one.

Let's count again.

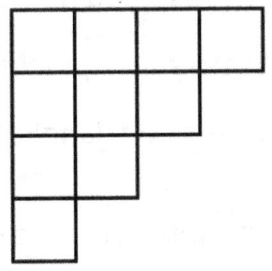

Size	#	Size	#	Size	#	Size	#
1×1	10	2×1	6	3×1	3	4×1	1
1×2	6	2×2	3	3×2	1		
1×3	3	2×3	1				
1×4	1						

Total = 1 + (1 + 3) + (1 + 3 + 6) + (1 + 3 + 6 + 10) = 1 + 4 + 10 + 20 = 35. This is the tetrahedral number $\binom{7}{4}$. Now what Polya?

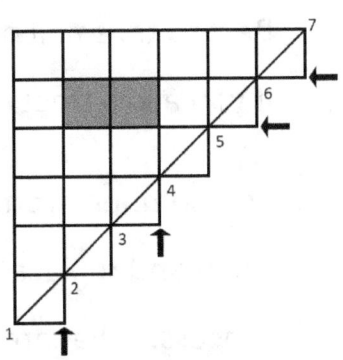

Considerable effort (*Mathematics Teacher*, May, 2010) went into this slick, easy to generalize, combinatorial proof. Embed the 4-tableau shape into a larger 6-tableau shape, draw the line as shown, and number the lattice points where the line intersects the extended shape with the numbers 1, 2, 3, 4, 5, 6, 7. Any selection of 4 of these integers will determine a unique rectangle. Draw a pair of vertical lines through the leftmost 2 points, together with a pair of horizontal lines through the rightmost 2 points. The selection 2, 6, 4, 5 is shown.

Also, any of the 35 rectangles can be matched in a one-to-one manner with a particular choice of 4 of these integers. In general, an n-tableau has $\binom{n+3}{4}$ rectangles.

More Gems For You

Here are some more gems for you to work on. Also included are some hints and partial solutions to get started.

1. Suppose you have a glass rod of length one meter, and you drop it, breaking it into three pieces. What is the probability that these three pieces form the sides of a triangle?

2. Find integers $a_1, a_2, a_3, \ldots, a_n$ and k such that $a_1 + a_2 + \ldots + a_n = k$ and the product $a_1 a_2 a_3 \ldots a_n$ is as large as possible. As an example, for k = 10, 5 + 5 = 3 + 7 = 1 + 2 + 3 + 4 = 10 are all good guesses but do not yield the maximum. Best to try small numbers like 10, 20, 50, and then 1000 first. What happens if you allow fractions? How about real numbers?

3. The numbers from 1 to 100 are written on the blackboard. Erase any two of them and replace them with their sum plus their product. Continue this process. What are the possible outcomes? Here is an example: erase 3 and 8 and replace them with 3 + 8 + 24 = 35. You now have 99 numbers left. (It is fine if there are two 35's at some point.)

4. Prove that $n^2 + 3n + 5$ is never divisible by 121 for any natural number n.

5. Show that every natural number greater than 11 is a sum of two composite numbers. Notice that you cannot do 11, but 12 = 4 + 8, 13 = 4 + 9, ...

Hints and Partial Solutions

1. Approach this one geometrically. Pick a point inside any equilateral triangle. Now draw perpendiculars to the three sides of the triangle. What is the total length of these three perpendiculars? See picture.

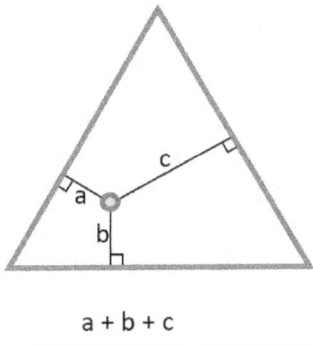
a + b + c

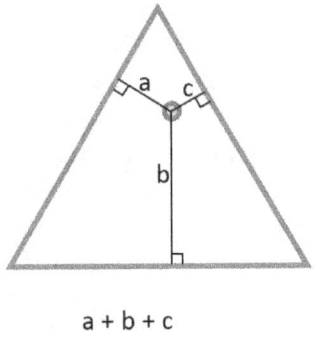
a + b + c

2. You never need any 9's, or 8's, or 7's, or... as summands. You can replace 9 with 3+ 3 +3 which has 27 as product; replace 8 with two 4's or 3 + 3 + 2; replace 7 with 3 + 4 and so on. When you tried $\frac{5}{2}+\frac{5}{2}+\frac{5}{2}+\frac{5}{2} = 10$, experimenting with fractions, $\frac{625}{16} > 39$, which was bigger than the product you decided on when using integers. Notice how close 2.5 is to a familiar constant.

3. Big hint: a + b + ab = (a + 1)(b + 1) – 1

4. Try to write a program and collect data; $n^2 + 3n + 5$ is always odd.

5. Write a program.

Chapter 3 – Combinatorial Proofs

A combinatorial proof is sometimes used to show that two very different looking expressions are in fact equal. The technique is as follows. Refer to the two different looking expressions as the left-hand side (LHS) and the right-hand side (RHS). Create a situation or question that is answered by the LHS, and then show that the RHS also answers the question. The conclusion is that LHS=RHS.

In the following we present a number of theorems, statements, and identities, giving combinatorial proofs of each. For each result the reader is urged to attempt an alternate proof for comparison purposes. Such an alternate proof could be algebraic or geometric in nature; or one could try a collapsing sum or an induction proof.

THEOREM 1: $\binom{2n}{2} = 2\binom{n}{2} + n^2$

Split the 2n objects into two groups A and B as shown:

$$\underbrace{\cdots\cdots}_{A} \qquad \underbrace{\cdots\cdots}_{B}$$

First, you can choose 2 objects from a set of 2n objects in $\binom{2n}{2}$ ways. Alternatively, you could select two from group A in $\binom{n}{2}$ ways or two from group B in $\binom{n}{2}$ ways or take one from each in $n \cdot n = n^2$ ways. Now add $\binom{n}{2} + \binom{n}{2} + n \cdot n$ and the result follows. The reader should attempt an algebraic proof using the factorial formula for $\binom{n}{k}$.

57

THEOREM 2: $\binom{m+n}{2} - \binom{m}{2} - \binom{n}{2} = mn$

Suppose you have a group of m men and n women and you want to form men-women dancing pairs. This can clearly be done in mn ways. Or, you could choose 2 from the total of m + n and delete the man-man pairs (there are $\binom{m}{2}$ of these) and delete the woman-woman pairs (also $\binom{n}{2}$ of these). The result follows.

The reader could attempt an algebraic proof or perhaps a geometric proof making use of figures consisting of triangular numbers. Note that THEOREM 1 is a special case of this. Also, as a challenge, the reader could formulate a similar result involving $\binom{a+b+c}{3}$ and a corresponding proof.

THEOREM 3: $\binom{n}{0} + \binom{n}{1} + \binom{n}{2} + \ldots + \binom{n}{n} = 2^n$

Here again we first create a question that is answered by either side of the given identity. Question: How many subsets does $\{a_1, a_2, \ldots, a_n\}$ have? This n-set has 2^n subsets. The left-hand side counts these subsets by their size. There are $\binom{n}{k}$ subsets of size k.

In this situation the reader might *not* want to try this algebraically.

THEOREM 4: $\binom{n}{1} + 2\binom{n}{2} + 3\binom{n}{3} + \ldots + n\binom{n}{n} = n2^{n-1}$

Given a set of n people, we can select a committee of size k along with a chair from that committee in $k\binom{n}{k}$ ways. We can select a committee (of size 1, or size 2, or ...) and its chair in $\binom{n}{1} + 2\binom{n}{2} + 3\binom{n}{3} + \ldots + n\binom{n}{n}$ ways. Alternatively, we can explain the term $n2^{n-1}$ as follows: choose one of the n people to chair any of the 2^{n-1} subsets of the remaining $n-1$ people.

The reader is invited to investigate one or more of the following approaches: $n2^{n-1}$ looks like a derivative, so try differentiating $(1+x)^n$; a reverse and add approach also works; or, first prove $k\binom{n}{k} = n\binom{n-1}{k-1}$ and then use it.

THEOREM 5: $\binom{n}{k} = \binom{n-1}{k} + \binom{n-1}{k-1}$

$\binom{n}{k}$ is the number of subsets of $\{a_1, a_2, a_3, \ldots, a_n\}$ of size k. Now a subset A of size k either contains the fixed element a_i or it does not. If A contains a_i, the remaining $k-1$ elements can be selected in $\binom{n-1}{k-1}$ ways. If, on the other hand, A does not contain a_i, you can choose the k elements from the depressed set $\{a_1, a_2, \ldots, a_{i-1}, a_{i+1}, \ldots, a_n\}$ in $\binom{n-1}{k}$ ways. Since these two cases are mutually exclusive, the theorem follows.

Once again, the reader is invited to attempt an algebraic proof.

THEOREM 6: Let d_n denote the number of derangements of 1, 2, 3, ..., n with $d_0 = 1, d_1 = 0$. Then $d_n = (n-1)(d_{n-1} + d_{n-2})$ for $n \geq 2$.

In forming a derangement of 1, 2, 3, ..., n, the integer n can be placed in any of the n − 1 spots 1, 2, 3, ..., n − 1, say spot i. If i goes into spot n, there are d_{n-2} ways to finish it. If i does not go into spot n, there are d_{n-1} ways to complete the derangement.

The reader can use the principle of inclusion-exclusion to derive a formula for d_n from which the new recursion $d_n = nd_{n-1} + (-1)^n$ and the above recursion can be derived.

THEOREM 7: $\binom{n}{0}^2 + \binom{n}{1}^2 + \binom{n}{2}^2 + \dots + \binom{n}{n}^2 = \binom{2n}{n}$

Given a group of 2n people consisting of n men and n women, in how many ways can one choose a group of n people? The answer to that question is just $\binom{2n}{n}$, the right side of the identity in question. Alternatively, one could also form the group of n people in the following way: choose 0 men and n women in the following:

$\binom{n}{0}\binom{n}{n} = \binom{n}{0}^2$ ways, or choose 1 man and n − 1 women in

$\binom{n}{1}\binom{n}{n-1} = \binom{n}{1}^2$ ways, or choose 2 men and n − 2 women in

$\binom{n}{2}\binom{n}{n-2} = \binom{n}{2}^2$ ways and so on. Now add these disjoint cases.

An alternate algebraic proof is less interesting: Extract the coefficient of x^n from both sides of $[(x+1)^n]^2 = (x+1)^{2n}$.

THEOREM 8: $\binom{2}{2} + \binom{3}{2} + \binom{4}{2} + \ldots + \binom{n}{2} = \binom{n+1}{3}$

The term $\binom{n+1}{3}$ is the number of binary strings of length $n+1$ consisting of three 1's (and the rest 0's). The left-hand side counts these by where in the string the leftmost 1 appears. Let $a_1 a_2 a_3, \ldots, a_{n+1}$ be a string of length $n+1$. There are $\binom{n}{2}$ strings when $a_1 = 1$, $\binom{n-1}{2}$ strings when $a_2 = 1$ is the leftmost 1, ..., and $\binom{2}{2}$ strings when $a_{n-1} = 1$ is the leftmost 1. In this last case, the string looks like $000\ldots 0111$.

Attempting a proof by mathematical induction is an easy option. An algebraic approach is not!

THEOREM 9: The number of positive integers that have their digits in strictly increasing order is $2^9 - 1$. Include single digit numbers.

There are $\binom{9}{1}$ single digit type, $\binom{9}{2}$ double digit type (just select 2 of the 9 digits 1, 2, 3, ..., 9 and arrange in order), ..., and so on to see that there are $\binom{9}{9}$ nine digit type. The total is $\binom{9}{1} + \binom{9}{2} + \binom{9}{3} + \ldots + \binom{9}{9} = 2^9 - 1$.

Here is an alternative, more clever, proof. Any non-empty subset of $\{1, 2, 3, \ldots, 9\}$ corresponds to a unique increasing number. There are $2^9 - 1$ such subsets. For example, the subset $\{4, 2, 9, 7\}$ corresponds to 2,479. Combining these two approaches actually gives you a nice proof that $\binom{9}{0} + \binom{9}{1} + \binom{9}{2} + \ldots + \binom{9}{9} = 2^9$.

THEOREM 10: $\binom{3n}{3} = 3\binom{n}{3} + 6n\binom{n}{2} + n^3$

This one is a little tougher. First rewrite as $n^3 = \binom{3n}{3} - 3\binom{n}{3} - 6n\binom{n}{2}$. Suppose you have n men, n women and n children and you want to select triples consisting of one man, one woman and one child. There are n^3 ways to do this, just pick one from each group. Alternatively, select 3 of the 3n people in $\binom{3n}{3}$ ways and delete the "bad" ones. Delete the ones where you selected all three from one group – there are $3\binom{n}{3}$ of these. Now, delete those where you had two from one group and one from another– there are $2n\binom{n}{2} + 2n\binom{n}{2} + 2n\binom{n}{2}$ of these.

THEOREM 11: $1 \cdot 1! + 2 \cdot 2! + 3 \cdot 3! + \ldots + n \cdot n! = (n+1)! - 1$

In how many ways can you arrange the n+1 numbers 0, 1, 2, …, n so that they are *not* in ascending order? The answer is $(n+1)! - 1$ since 0, 1, 2, …, n is the *only* arrangement in ascending order. Now, let's separate into cases. Let $a_0, a_1, a_2, \ldots, a_n$ represent an arrangement of these n+1 numbers. If $a_0 \neq 0$, there are n choices left for a_0, and then n! ways to fill out $a_1, a_2, \ldots, a_n$ for a total of $n \cdot n!$. Now let $a_0 = 0$ but $a_1 \neq 1$. There are $n-1$ choices for a_1 and $(n-1)!$ ways to complete for a total of $(n-1)(n-1)!$. Now continue with $a_0 = 1, a_1 = 1$ but $a_2 \neq 2$. There are $(n-2)(n-2)!$ ways, and so on.

The reader should attempt a collapsing sum or induction proof.

THEOREM 12: $1 \cdot n + 2(n-1) + 3(n-2) + \ldots + n \cdot 1 = \binom{n+2}{3}$

Let $S = \{0, 1, 2, \ldots, n, n+1\}$. The number of subsets of S of size 3 is $\binom{n+2}{3}$. Each one looks like $\{a, b, c\}$ with $a < b < c$. Let's count these by looking at the size of the middle element b. If b=1, there is one choice for a, namely a=0 and n choices for c for a total of $1 \cdot n$. If b=2, there are 2 choices for a and $n-1$ choices for c for a total of $2(n-1)$. If b=3, the total is $3(n-2)$, and so on. The total derived by looking at cases is $1 \cdot n + 2(n-1) + 3(n-2) + \ldots + n \cdot 1$ and this must equal $\binom{n+2}{3}$ since the cases are disjoint.

THEOREM 13: $k\binom{n}{k} = n\binom{n-1}{k-1}$

Suppose you have a group of n people and you wish to form a subcommittee of k people with one of those k people to serve as chair. Choose the subcommittee in $\binom{n}{k}$ ways and the chair in k ways. The product rule gives $k\binom{n}{k}$ as the number of ways of selecting such a chaired subcommittee.

Alternatively, you could first choose any one of the n people to serve as chair and then fill out the committee in $\binom{n-1}{k-1}$ ways. There are $n\binom{n-1}{k-1}$ ways to select a chaired subcommittee. Hence $k\binom{n}{k} = n\binom{n-1}{k-1}$. This can be used to provide an alternate proof of THEOREM 4.

The reader should attempt the easier algebraic technique by converting $\binom{n}{k}$ to factorial form.

THEOREM 14: $P(n, k) = k! \binom{n}{k}$

Question-- How many permutations are there of k objects chosen from a collection of n objects? The LHS answers the question. There are $P(n, k) = n(n - 1)(n - 2)\ldots(n - k + 1)$ ways. Alternatively, one could first choose the k objects from the n objects in $\binom{n}{k}$ ways and then permute these in k! ways.

THEOREM 15: $n2^{n-1} = 1\binom{n}{1} + 2\binom{n}{2} + 3\binom{n}{3} + \ldots + n\binom{n}{n}$

Contrast this discussion with that presented in THEOREM 4. Look at the set of the first 2^n nonnegative integers $0, 1, 2, \ldots, 2^n - 1$. When you convert each to binary form what is the total number of 1s written? This binary list will look like the standard listing in B^n the set of all binary strings of length n. For n=3, $B^3 = \{000, 001, 010, 011, 100, 101, 110, 111\}$. There are $3 \cdot 2^2$ in B^3. In B^n, each string has length n and there are 2^n of them. But, half of the $n \cdot 2^n$ symbols are 1's. Then, the total number of 1's is $n2^{n-1}$. Alternatively, we could consider each string and count those with one 1, then those with two 1's, etc. There are $1 \cdot \binom{n}{1}$ with one 1, $2\binom{n}{2}$ total 1's in those binary numbers with exactly two 1's, $3\binom{n}{3}$ in those with exactly three 1's, and so on. The total is $1\binom{n}{1} + 2\binom{n}{2} + 3\binom{n}{3} + \ldots + n\binom{n}{n}$. The result now follows by equating $n2^{n-1}$ to this sum.

THEOREM 16: $\binom{n}{0}^2 + \binom{n}{1}^2 + \binom{n}{2}^2 + \ldots + \binom{n}{n}^2 = \binom{2n}{n}$

Let's revisit this identity using equivalence relations. A binary relation R on the set of all binary strings of length n is defined by specifying that $(\alpha, \beta) \in R$ whenever weight α = weight β. This R is an equivalence relation. For n=3 there are four different equivalence classes, each containing strings of weight 0, 1, 2, or 3. The relation R contains $1^2 + 3^2 + 3^2 + 1^2$ ordered pairs; for example, with weight 1 each of the three strings 001, 010, 100 can be paired with any one of those same strings for a total of $3^2 = 9$. These ordered pairs can be counted in another way. Each ordered pair looks like $(---,---)$. Place 1's in any three of the six positions, and 0's in the others. If you take the complement of the entries in the second coordinate an element of R is produced. Here is what one sequence of this process looks like:

$(---,---) \to (-11,--1) \to (011, 001) \to (011, 110)$.

The reader can check that this process always produces an element of R and that the case of n=3 extends easily to general n. Conclusion: choose the n positions for 1's in $\binom{2n}{n}$ ways. The result follows.

THEOREM 17: $\binom{n}{0}d_0 + \binom{n}{1}d_1 + \binom{n}{2}d_2 + \ldots + \binom{n}{n}d_n = n!$ where d_n denotes the nth derangement number, $d_0 = 1, d_1 = 0$.

The right-hand side, $n!$, gives the number of permutations of n objects. So, the left-hand side must provide the same enumeration. The left side partitions the permutations according to how many elements are deranged (and the rest fixed). The term $\binom{n}{i}d_i = \binom{n}{n-i}d_i$ gives the number of permutations of n where $n-i$ elements are fixed and the remaining i elements are deranged. Summing over all i yields the following:

$$\binom{n}{n}d_0 + \binom{n}{n-1}d_1 + \ldots + \binom{n}{0}d_n = \binom{n}{0}d_0 + \binom{n}{1}d_1 + \ldots + \binom{n}{n}d_n = n!$$

THEOREM 18: $F_{n+1} = \binom{n}{0} + \binom{n-1}{1} + \binom{n-2}{2} + \ldots$ where F_n denotes the nth Fibonacci number.

Here is a question that might resolve the issue: how many different brick paths of length n (and width 1) can you make using 1×1 bricks and 1×2 bricks? Let a(n) denote the number of such paths of length n. A few drawings will show that a(1)=1, a(2)=2, a(3)=3, a(4)=5. Since you can place a 1×1 brick in front of all paths of length $n - 1$ or a 1×2 brick in front of all paths of length $n - 2$ we have that $a(n) = a(n-1) + a(n-2)$. This recursion, along with the initial conditions, shows that $a(n) = F_{n+1}$, the left-hand side of the identity.

Now let's look at all paths of length n and count them by the number of 1×2 bricks. If there are i 1×2 bricks, there are $n - i$ total bricks making up the path of length n. Choose the positions of the i 1×2 bricks in $\binom{n-i}{i}$ ways. Now sum as i ranges through the values

0, 1, 2, ..., and obtain $\binom{n}{0} + \binom{n-1}{1} + \binom{n-2}{2} + \ldots = F_{n+1}$.

The reader should draw all paths of length n = 5, for example, and examine the cases with i=0, 1, 2. The reader could also explore other proofs.

<u>THEOREM 19</u>: $\binom{m+n}{2} - \binom{m}{2} - \binom{n}{2} = mn$ (Revisited)

1.

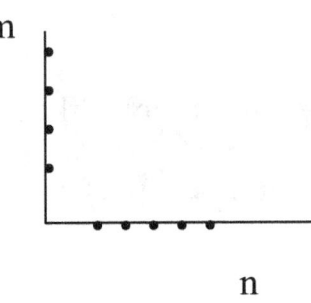

How many lines that are not vertical or horizontal can you form by connecting the points that are on the positive x and y axes? Since we really want only lines that are formed by connecting the m points with the n points, we could say we have mn lines. Or we could consider all m + n points and pick 2 in $\binom{m+n}{2}$ ways and then delete those choices where you took 2 from m or 2 from n, since these formed vertical and horizontal lines, resp.; then,

$$\binom{m+n}{2} - \binom{m}{2} - \binom{n}{2} = mn.$$

2. There are mn 1×1 squares in the subdivided m by n rectangle. Each choice of arrows (one horizontal, one vertical) specifies one of these squares. Pick two arrows but don't take two from the top or two from the side. Then $\binom{m+n}{2} - \binom{m}{2} - \binom{n}{2} = mn$

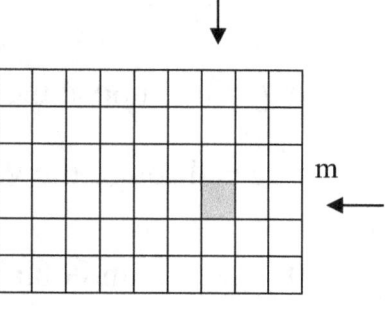

3. Take m people in one group and n in another. How many handshakes can be accomplished? Among the m people, there are $\binom{m}{2}$ handshakes; among the n people, there are $\binom{n}{2}$ handshakes and between the two groups, mn. But, $\binom{m+n}{2}$ also represents the total number of handshakes among the m + n people. Then we get: $\binom{m}{2} + \binom{n}{2} + mn = \binom{m+n}{2}$.

4. Here is another approach. From $\binom{m+n}{2} = \binom{m}{2} + \binom{n}{2} + mn$, it looks like a large triangular number is made up of two smaller triangular numbers plus a rectangle. Try this with m = 5 and n = 3.

$$\binom{8}{2} = \binom{5}{2} + \binom{3}{2} + 15$$

Now build it!

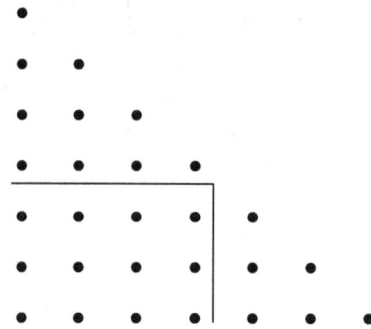

68

THEOREM 20: $1 \cdot \binom{n}{1}^2 + 2 \cdot \binom{n}{2}^2 + 3 \cdot \binom{n}{3}^2 + \cdots + n \cdot \binom{n}{n}^2 = n\binom{2n-1}{n-1}$

A mathematics team is chosen from a group of n men and n women. The team will consist of n people with a woman to be selected as captain. This can be done in $n\binom{2n-1}{n-1}$ ways; select one of the n women first, then the other n-1 can be selected from the remaining 2n–1 people. Alternatively, you can choose a team with:

1 woman and n–1 men in $\binom{n}{1}\binom{n}{n-1} = \binom{n}{1}^2$ ways with 1 choice for captain

2 women and n–2 men in $\binom{n}{2}\binom{n}{n-2} = \binom{n}{2}^2$ ways with 2 choices for captain

. . .

n woman and 0 men in $\binom{n}{n}\binom{n}{0} = \binom{n}{n}^2$ ways with n choices for captain.

THEOREM 21:

This one is not so much a theorem, but rather an identity between two expressions involving binomial coefficients. Count, in two different ways, the number of triangles you can draw using the vertices below and show that:

$$5\binom{7}{2} + \binom{5}{2}\binom{7}{1} + \binom{5}{3} = \binom{12}{3} - \binom{7}{3}$$

.

. .

. .

. . . .

For the RHS choose any 3 of the 12 pts and delete those where 3 came from the bottom line. For the LHS choose 2 from the line and any one from the others, or choose 1 from the line of 7 points, along with 2 from the 5; or 3 from the top 5.

Chapter 4 – Visual Approach

Not all argument are strict, traditional, proofs that need to contain variables, steps, and all the usual mathematical components. A proof is a convincing argument for whatever audience that is present, elementary school to graduate students. Giving a convincing argument will be different, depending on the audience makeup. At times, a visual approach is most appropriate. This section illustrates how problems can be analyzed visually. We are hoping that from a fairly simple diagram, you will be able to imagine a complex mathematical statement. Let's start with an example.

What do you see when you look at this diagram? A walkway around a park? A subdivided square? A land-surveyor's view of a future apartment complex?

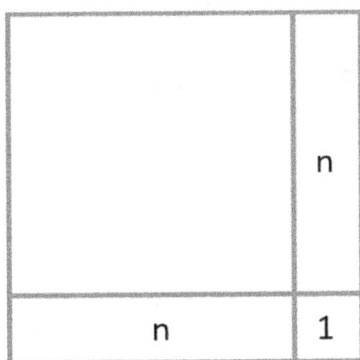

Actually, it could be a picture "proof" of the statement "every odd number can be written as a difference of two consecutive squares." Here is the algebra:

$(n+1)^2 - n^2 = n^2 + 2n + 1 - n^2 = 2n+1$

This next introductory example shows how one might unravel a complex identity using a combination of pictures.

(De)Compose Yourself

Let's see how we can make use of pictures 9 and 14 in handling a pretty tough problem. We do not want you to panic when faced with seeking a proof for identities like

$$1 \cdot n + 3(n-1) + 5(n-2) + \ldots + (2n-1) \cdot 1 = 1^2 + 2^2 + 3^2 + \ldots + n^2$$

First let's see what this identity is all about for a particular example. Let's try n = 5.

$$1 \cdot 5 + 3 \cdot 4 + 5 \cdot 3 + 7 \cdot 2 + 9 \cdot 1 = 1^2 + 2^2 + 3^2 + 4^2 + 5^2$$

This looks like a dot product of 1, 3, 5, 7, 9 with 5, 4, 3, 2, 1 (odd numbers with integers backwards). If you say this aloud reading from right to left – one 9, two 7's, three 5's, four 3's and five 1's --you can decompose this dot product to look like:

$$9 \ 7 \ 5 \ 3 \ 1 = 5^2$$
$$7 \ 5 \ 3 \ 1 = 4^2$$
$$5 \ 3 \ 1 = 3^2$$
$$3 \ 1 = 2^2$$
$$1 = 1^2$$

where we are making use of picture #9 (below)

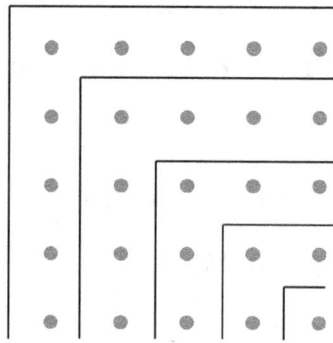

72

showing that the sum of consecutive odd numbers is a perfect square; in this case $9 + 7 + 5 + 3 + 1 = 5^2$. Finally, add the six columns above, starting with the 9, then $7 + 7$, then $5 + 5 + 5$, then $3 + 3 + 3 + 3$, then finally $1 + 1 + 1 + 1 + 1$.

Here is a different approach. If you view the original problem backwards like

$$(2n-1)\cdot 1 + (2n-3)\cdot 2 + \ldots + 1\cdot n = 1^2 + 2^2 + \ldots + n^2$$

the following decomposition (which is picture #14), with $n = 5$, sheds a different light on a general proof.

```
1 1 1 1 1       1 1 1 1 1
1 2 2 2 2       1 1 1 1 1      1 1 1 1 1
1 2 3 3 3  =    1 1 1 1 1  +   1 1 1 1 1   +   1 1 1   +   1 1   + 1
1 2 3 4 4       1 1 1 1 1      1 1 1 1 1       1 1 1       1 1
1 2 3 4 5       1 1 1 1 1      1 1 1 1 1       1 1 1
                               1 1 1 1 1
```

On the left side we have 9 1's, 7 2's, 5 3's, 3 4's and one 5 and on the right side we have the desired number of squares – $25 + 16 + 9 + 4 + 1$.

Poster Instructions – Matching

Match each of the following identities with the appropriate diagrams on the next pages by placing a number in each space. You also might practice saying aloud what each identity is about.

____ A. $\binom{n}{2} + \binom{n+1}{2} = n^2$

____ B. $1 + 3 + 5 + 7 + \cdots + 2n - 1 = n^2$

____ C. $1 + 2 + 3 + 4 + \cdots + n = \frac{n(n+1)}{2}$

____ D. $\binom{n}{0}^2 + \binom{n}{1}^2 + \binom{n}{2}^2 + \cdots + \binom{n}{n}^2 = \binom{2n}{n}$

____ E. $F_1^2 + F_2^2 + F_3^2 + \cdots + F_n^2 = F_n F_{n+1}$

____ F. $\frac{1}{2} + \frac{1}{4} + \frac{1}{8} + \frac{1}{16} + \cdots = 1$

____ G. $1 + 2 + 3 + \cdots + n + (n-1) + \cdots + 3 + 2 + 1 = n^2$

____ H. $1 + 3 + 5 + \cdots + (2n-1) = 1 + 2 + 3 + \cdots + n + \cdots + 3 + 2 + 1$

____ I. $1^2 + 2^2 + 3^2 + \cdots + n^2 = \binom{n+1}{3} + \binom{n+2}{3}$

____ J. $(2n-1) + \binom{n}{2} + \binom{n-1}{2} = n^2$

____ K. $1(n) + 3(n-1) + 5(n-2) + \cdots + (2n-1)1 = 1^2 + 2^2 + 3^2 + \cdots + n^2$

____ L. $\binom{m+n}{2} - \binom{m}{2} - \binom{n}{2} = mn$

____ M. $\frac{1}{4} + \frac{1}{4^2} + \frac{1}{4^3} + \frac{1}{4^4} + \cdots = \frac{1}{3}$

____ N. $1 + 3 + 6 + 10 + \cdots + \frac{n(n+1)}{2} = \binom{n+2}{3}$

____ O. $2 + 4 + 6 + 8 + \cdots + 2n = n(n+1)$

1.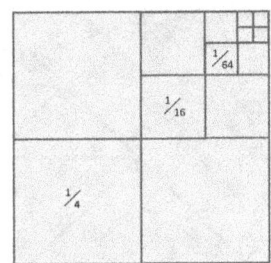

2.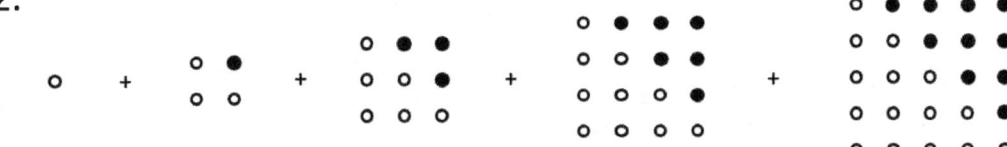

3.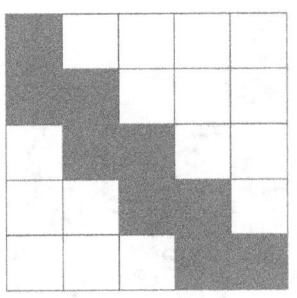

4.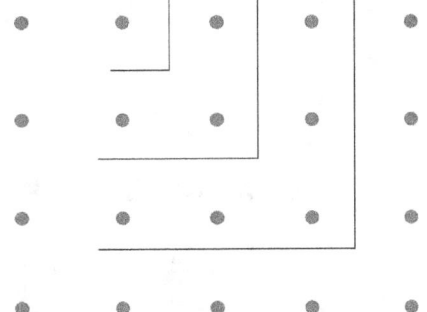

5.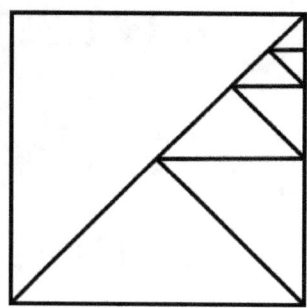

6.

7.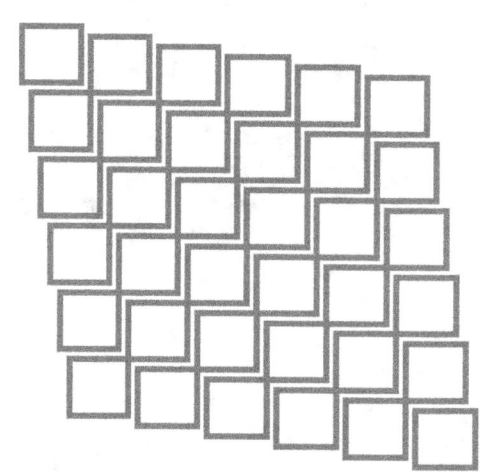

8.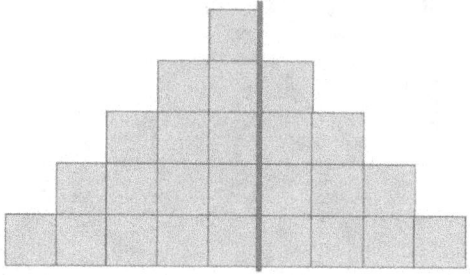

9.

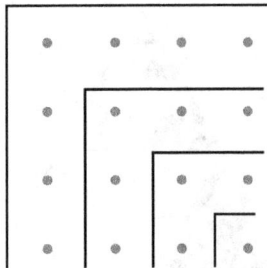

10.

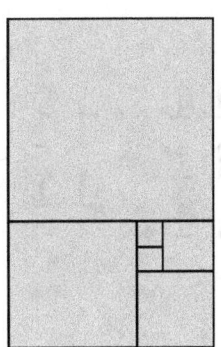

11.

12.

13.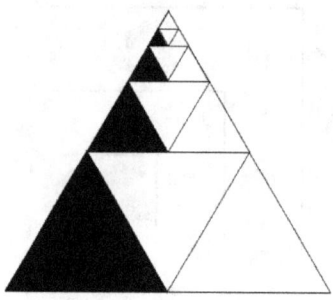

14.
$$\begin{matrix}1&1&1&1\\1&1&1&1\\1&1&1&1\\1&1&1&1\end{matrix} + \begin{matrix}1&1&1\\1&1&1\\1&1&1\\1&1&1\end{matrix} + \begin{matrix}1&1\\1&1\\1&1\end{matrix} + 1 = \begin{matrix}1&1&1&1\\1&2&2&2\\1&2&3&3\\1&2&3&4\end{matrix}$$

15.

```
        N
        1
       1 1
      1 2 1
     1 3 3 1
      4 6 4
      10 10
       20
        S
```

16.

```
o
o o
o o o
o o o o
o o o o o |
o o o o o | o
o o o o o | o o
```

Verbal Descriptions

One current concern in mathematics education is the lack of opportunity for students to "talk mathematics", to verbalize mathematical thoughts and ideas. Such rigidity stunts mathematical growth, and in some cases is the cause of statements like "I hate mathematics".

Here you have a chance to practice "talking mathematics". Say these 15 statements aloud, try to understand what you are actually saying, and try to form a mental image of the idea. And then match them to the identities A – O and then to their diagrams.

____ Add up a bunch of squares to form a rectangle.

____ A tetrahedral number is a sum of triangular numbers.

____ The sum of consecutive odd numbers is a square.

____ Make men – women dancing pairs.

____ Subdivide an equilateral triangle into 4 equal parts.

____ The sum of two consecutive triangular numbers is a perfect square.

____ Hey Gauss! Can two staircases make a rectangle?

____ Up to the middle and then back down the square.

____ The sum of two tetrahedral numbers is also the sum of a bunch of squares.

____ Symmetry along the horizon; right down the middle.

____ Share that one cookie; a half to a friend, a half of that to another friend,….

____ Oblong numbers combine to form a rectangle.

____ Start with a square. Remove 2 triangular numbers and see what is left.

____ Decompose a square several ways: Up and down is quite odd.

____ A mountain of level squares with a corner peak.

Chapter 5 – Computer Analysis

Despite common knowledge, mathematicians and computer scientists are quite different. There is plenty of overlap, but the fundamental differences will astound you. Whenever looking at a problem, determine if programming would be a suitable way to prove (or disprove) a theory. Programming may or may not be sufficient to grant satisfactory results. Even if the results are not satisfactory, it can give insight on the mathematical solution.

Confirming a proof to be true may be too steeped in mathematical theory for a computer program to solve. But a computer is most excellent at finding counterexamples and for producing data. If the correctness of a theory is under question, a computer program can be used to locate potential counterexamples, if any exist. It should be noted that this is only useful to disprove a theorem. To satisfy mathematicians, the program would have to work for all possible cases, which is far more difficult than finding a single example to disprove a theory.

Proof Using Code

As an example, the program below demonstrates the classic theorem of $1+2+3+\ldots+n = n(n+1)/2$, by calculating both sides of the equation and comparing results. The program can be run indefinitely and never find a counterexample to the formula. This template can be used to test similar equality problems.

Is this proof? A classic mathematician would say no, but an algorithmic mathematician may claim the program can be analyzed in the same manner as a mathematical proof. Whether or not a computer program can serve as a mathematical proof, it can still provide useful information.

As you read through this set of upcoming problems, our guess is that your initial response would be "those look tough, and I will probably need to gather significant data to guess at what is happening." That might be your first clue that some sort of technology or computer analysis would be helpful.

The following is a Python program to test the most basic of formulas. It can be used as a template to solve any problem, by changing the left and right side expressions. The `leftSide` function definition calculates the left hand side of the equation by adding all numbers requested to a total, thus requiring n additions to complete. Whereas the `rightSide` function definition calculates the formula `n(n+1)/2`, requiring only an addition, a multiplication, and a division. The other two function definitions s simply create formatted reports summarizing the two as equal.

But first, the template:

```python
def leftSide(n):
    "Calculates the left side of the equation"
    total = 0
    for i in range(1,n+1):
        total += i
    return total

def rightSide(n):
    "Calculates the right side of the equation"
    total = n*(n+1)/2
    return total

def testEqualityVerbose(startAt, upTo):
    "Tests all numbers, from startAt to upTo"
    print("Verbose Mode Testing")
    print("Testing equality for all numbers from", startAt, "to", upTo, "(verbose mode)")
    totalEqual = 0
    totalNotEqual = 0
    for i in range(startAt,upTo+1):
        left = leftSide(i)
        right = rightSide(i)
        if left == right:
            message = "equal"
            totalEqual += 1
```

```python
        else:
            message = "not equal"
            totalNotEqual += 1
        print("%4d   %6d   %6d   %s" % (i, left, right, message))
    print("There were",totalEqual,"lines correct.")
    print("There were",totalNotEqual,"lines not correct.")

def testEqualitySummary(startAt, upTo, progressSkipValue,
progressUpdateValue):
    "Tests all numbers, only prints not-equal, prints a # every
skipValue times"
    print("Summary Mode Testing")
    print("Testing equality for all numbers
from",startAt,"to",upTo,"(summary mode)")
    print("Each # is",progressSkipValue,"numbers tested.")
    print("Printing n every",progressUpdateValue,"iterations.")
    totalEqual = 0
    totalNotEqual = 0
    for i in range(startAt,upTo+1):
        left = leftSide(i)
        right = rightSide(i)
        if left != right:
            print("%4d   %6d   %6d   notEqual" % (i, left, right))
            totalNotEqual += 1
        else:
            totalEqual += 1
        if i%progressUpdateValue == 0:
            print("   n =",i)
        if i%progressSkipValue == 0:
            print("#",end="")
    print()
    print("There were",totalEqual,"lines correct.")
    print("There were",totalNotEqual,"lines not correct.")

def main():
    testEqualityVerbose(0, 25)
    testEqualitySummary(0, 10000, 1000, 10000)

main()
```

The output of the above program is as follows:

```
Verbose Mode Testing
Testing equality for all numbers from 0 to 25 (verbose mode)
   0        0        0   equal
   1        1        1   equal
   2        3        3   equal
   3        6        6   equal
   4       10       10   equal
   5       15       15   equal
   6       21       21   equal
```

```
 7     28    28  equal
 8     36    36  equal
 9     45    45  equal
10     55    55  equal
11     66    66  equal
12     78    78  equal
13     91    91  equal
14    105   105  equal
15    120   120  equal
16    136   136  equal
17    153   153  equal
18    171   171  equal
19    190   190  equal
20    210   210  equal
21    231   231  equal
22    253   253  equal
23    276   276  equal
24    300   300  equal
25    325   325  equal
There were 26 lines correct.
There were 0 lines not correct.
Summary Mode Testing
Testing equality for all numbers from 0 to 10000 (summary mode)
Each # is 1000 numbers tested.
Printing n every 10000 iterations.
  n = 0
##########   n = 10000
#
There were 10001 lines correct.
There were 0 lines not correct.
```

Conclusion? The above program tested the first 10000 natural numbers, and it worked for all of them. A simple change of n to 100,000 or 10,000,000 would have given similar results. The question remains: is this proof? Unfortunately, at this point, it is not. It is very strong evidence, but even the most convincing statements of evidence is still not mathematical proof.

Proof Using Existing Software

The next question will attract bipartisan (maybe even tripartisan) reactions. The question, simply stated, is to factor $x^5 + x + 1$.

This question was posed in a recent problem-solving course for secondary teachers. Several bright and creative participants came up with the following:

$$x^5 + x + 1 = (x^5 + x^4 + x^3) - (x^4 + x^3 + x^2) + (x^2 + x + 1)$$

$$= x^3(x^2 + x + 1) - x^2(x^2 + x + 1) + (x^2 + x + 1)$$

$$= (x^2 + x + 1)(x^3 - x^2 + 1)$$

This clever add-in subtract-out, followed with factoring by grouping, is borderline artistic.

There are two approaches to the problem. Grassl, a mathematician, would take an algebraic approach to the solution. However, Zeller would take a different approach. While he is familiar with rules of algebra, he is far more apt to use a computer program. The programming language Python may have a library or special function to do factoring, but something like this is not built-in. But there are other algorithmic solutions. Zeller's approach would be something along the following:

$$x^5 + x + 1 \quad = \text{go to } \text{http://WolframAlpha.com and enter x\textasciicircum 5 + x + 1}$$

$$= \text{Alternate forms: } (x^2 + x + 1)(x^3 - x^2 + 1)$$

Whose approach is better? The "mathematicians" in the audience would say that Grassl did it correctly, showing knowledge of algebra and expression. But Zeller's approach was not only easier, but it gave entirely more information in less time, requiring less mathematical knowledge. However, it had that tricky requirement of having the Internet available, and to live in a reality where Wolfram Alpha exists. Even 12 years ago, Zeller's approach would not have been possible, as the software was not written until 2009.

But Zeller's approach is not without the necessity for mathematical knowledge. The interface for entering the formula must be clear. The result was not the first in the list, requiring scrolling through unrequested information. So maybe this is math, and maybe it isn't.

But remember, the hidden goal here is not necessarily to actually factor the expression $x^5 + x + 1$, but rather to create a design, an interesting technique, thereby achieving a pleasing solution.

Anyone can push a button Zeller!

Problems

The following problems are suggested for computer analysis. They were written with the idea of using a computer program or some other technology to solve.

1. For which n is $1! + 2! + 3! + \cdots + n!$ a perfect square?

2. If the 5040 permutations of 1, 2, 3, 4, 5, 6, 7, are listed in increasing numerical order from the smallest 1,234,567 to the largest 7,654,321, what is the 800th number?

3. Prove the following for $n \geq 2$: $\frac{1}{n+1} + \frac{1}{n+2} + \cdots + \frac{1}{2n} > \frac{13}{24}$. **(left to the reader)**

4. Is F_n, the nth Fibonacci number, always bigger than $\left(\frac{3}{2}\right)^{n-1}$?

5. The five digits a, b, c, d, e, of 55225 are such that $a = b = e$ and $c = d$. Also, $55225 = 235^2$. Find another positive integer m such that m^2 is also a 5-digit number abcde that satisfies $a = b = e$ and $c = d$.

6. Which whole numbers are equal to the sum of the factorials of their digits?

7. For which n is $1^2 + 2^2 + \cdots + n^2$ a perfect square? How about $\binom{2}{2} + \binom{3}{2} + \binom{4}{2} + \cdots + \binom{n}{2}$? **(left to the reader)**

8. For which n is $n! + 1$ a perfect square?

9. For which n is F_n a perfect square?

10. How many of $101, 10101, 1010101, \ldots$ are prime?

11. For which n is $1^3 - 2^3 + 3^3 - 4^3 + \cdots + (-1)^{n+1}n^3$ a perfect square? **(left to the reader)**

12. What happens when you try to prove $\frac{n-1}{n+1} < \frac{9}{10}$ by mathematical induction?

13. Here is a theorem that is not difficult to prove, using the contrapositive. If 2^n-1 is prime, then n is prime. However, is the converse true? The converse would be: If n is prime, then 2^n-1 is prime.

Solutions

Question 1 – Sum of Factorials a Perfect Square

For which n is $1! + 2! + 3! + \cdots + n!$ a perfect square? The program for this was relatively simple to create. Two answers were found: $n = 1$ and $n = 3$. Here is the output:

Output:
```
1 --> 1
It holds for i = 1
2 --> 3
3 --> 9
It holds for i = 3
4 --> 33
5 --> 153
6 --> 873
```

```
7  --> 5913
8  --> 46233
9  --> 409113
10 --> 4037913
11 --> 43954713
12 --> 522956313
13 --> 6749977113
14 --> 93928268313
15 --> 1401602636313
16 --> 22324392524313
17 --> 378011820620313
18 --> 6780385526348313
19 --> 128425485935180313
20 --> 2561327494111820313
21 --> 53652269665821260313
22 --> 1177652997443428940313
23 --> 27029669736328405580313
24 --> 647478071469567844940313
25 --> 16158688114800553828940313
26 --> 419450149241406189412940313
27 --> 11308319599659758350180940313
28 --> 316196664211373618851684940313
29 --> 9157958657951075573395300940313
30 --> 274410184701421342097037809403130313
It holds for i = 30
31 --> 8497249472648064951935266660940313
It holds for i = 31
32 --> 271628086406341595119153278820940313
It holds for i = 32
33 --> 8954945705218228090637347680100940313
It holds for i = 33
34 --> 304187744744822368938255957323620940313
It holds for i = 34
```

This data is striking! Why do all of these numbers end in 3, will that continue, and what is wrong with that? The sum up to 4! is 33. Additionally, when you add $5! = 120$ to 33, you will get a 3 in the units position, and so on. You can "smell" this proof a mile away. Now, no squares end in a 3 (they instead end in 0, 1, 4, 9, 6, 5, 6, 9, 4, 1). This is a perfect, classic example where the computer analysis does the heavy lifting but cannot complete the proof. The reader captures the data and renders the proof. The code that produces the output is as follows:

```
# Q1 -- For which n is 1! + 2! + 3! + ... + n! a perfect square?

import math

def isPerfectSquare(n):
    a = math.sqrt(n)
    b = int(math.sqrt(n))
    if a == b:
        return True
    return False

def sumOfFactorials(n):
    sum = 0
    for i in range(1,n+1):
        sum += math.factorial(i)
    return sum

def main():
    for i in range(1,35):
        print(i,"-->",sumOfFactorials(i))
        if isPerfectSquare(sumOfFactorials(i)):
            print("It holds for i =",i)

main()
```

Conclusion: In this case, limitations of the program storage space caused incorrect results. When the output is interpreted correctly, this program still serves as convincing evidence that the statement is true. But, as always, "evidence is not proof."

Question 2 – 800th permutation of 1234567

If the 5040 permutations of 1, 2, 3, 4, 5, 6, and 7 are listed in increasing numerical order from smallest (1234567) to largest (7654321), what is the 800th number? Not all programming solutions are lengthy. The best and most elegant solutions are those that make use of existing commands and libraries that fit to the task. In this case, the permutations command imported from the `itertools` library does the purpose. The permutations command creates all permutations of the parameter in order, and the 800th number from the list is printed.

```
from itertools import permutations
perm = permutations([1, 2, 3, 4, 5, 6, 7])
print(list(perm)[799])
```

The resulting output is:

(2, 1, 6, 4, 3, 7, 5)

By hand, this problem is considerably more tedious and prone to errors. However, knowing that 6! = 720 is quite helpful for doing it by hand.

Question 4 – F$_n$ is always bigger than $\left(\frac{3}{2}\right)^{n-1}$

Is F$_n$, the nth Fibonacci number, always bigger than $\left(\frac{3}{2}\right)^{n-1}$? An official definition of "always" usually contains an upside-down A meaning "for all."

The answer to the question, as shown by counterexample, is no, there are at least four cases where the nth Fibonacci number is smaller than $\left(\frac{3}{2}\right)^{n-1}$. The output from the problem shows the cases:

```
       fibon    (3/2)^(n-1)
           1           1.0
           1           1.5   fibon smaller
           2           2.2   fibon smaller
           3           3.4   fibon smaller
           5           5.1   fibon smaller
           8           7.6
          13          11.4
          21          17.1
```

When the program is continued, no further occurrences result. A truly industrious mathematician could prove that there will be no more. However, this does answer the question with a clear negative. The code to produce the output above is as follows:

```
#Q4 -- Is Fn, the nth Fibonacci number, always bigger than
#(3/2)^(n-1)?

def fibonacci(n):
    if n<0:
        print("Warning -- Incorrect input")
    elif n==1:
        return 1
    elif n==2:
        return 1
    else:
        return fibonacci(n-1)+fibonacci(n-2)


def rightSide(n):
    "Calculates the right side of the equation"
    return (3/2)**(n-1)


def main():
    print("%12s %12s" % ("fibon","(3/2)^(n-1)"))
    for i in range(1,100):
        fibon = fibonacci(i)
        right = rightSide(i)
        print("%12d %12.1f" % (fibon,right), end="")
        if right>fibon:
            print("   fibon smaller")
        else:
            print()

main()
```

So no, F_n is not always larger than $(3/2)^{n-1}$. This program proves that without a doubt by finding four counterexamples. The problem statement could be adjusted, since all counterexamples were less than 10. But that is a different problem.

Question 5 – m^2 in XXYYX form

The five digits a,b,c,d,e of 55225 are such that a = b = e and c = d. Also, $55225 = 235^2$. Find another positive integer m such that m^2 is also a 5-digit number abcde that satisfies a = b = e and c = d. There are times one does not want the computer program to do everything. In this example, the square root of every possible combination of numbers is given, and it is up to the user to determine if it is a perfect square or not. This reduces the possibility that there is something wrong with the perfect square algorithm. It may be necessary to automate the perfect square algorithm for other problems, but in this example a human can just as easily scan the output for the results.

```
#Q5 -- The five digits a,b,c,d,e of 55225 are such that
#a = b = e and c = d.  Also, 55225 = 235^2.  Find another
#positive integer m such that m2 is also a 5-digit number
#abcde that satisfies a = b = e and c = d.
from math import sqrt
#there are only 100 lines of output.  Find the two lines
#that result in integers
def main():
    for abe in range(10):
        for cd in range(10):
            num = abe*10000 + abe*1000 + cd*100 + cd*10 + abe
            print(num, sqrt(num))

main()
```

The resulting output is below. Just find the lines that are perfect squares. All other lines resulted in non-integers.

0 0.0	33443 182.87427375112117	66886 258.62327814796566
110 10.488088481701515	33553 183.17477992343814	66996 258.8358553214759
220 14.832396974191326	33663 183.4747939091362	77007 277.501351348061
330 18.16590212458495	33773 183.77431811871864	77117 277.69947785330817
440 20.97617696340303	33883 184.07335494307696	77227 277.89746310464943
550 23.45207879911715	33993 184.37190675371343	77337 278.09530740377477
660 25.69046515733026	44004 209.77130404323657	77447 278.29301105130185
770 27.748873851023216	44114 210.03333068825052	77557 278.4905743467811
880 29.664793948382652	44224 210.29503084951864	77667 278.6879975887013
990 31.464265445104548	44334 210.55640574439906	77777 278.88528107449486
11001 104.88565202161828	44444 210.81745658270333	77887 279.08242510054265
11111 105.40872829135166	44554 211.0781845667619	77997 279.2794299621796
11221 105.92922165295089	44664 211.33859089148862	88008 296.6614231746352
11331 106.44716999526104	44774 211.5986767444447	88118 296.8467618149135
11441 106.96261028976434	44884 211.85844330590177	88228 297.0319848097171
11551 107.47557862137798	44994 212.11789174890458	88338 297.21709237525357
11661 107.98611021793498	55005 234.53144778472674	88448 297.4020847270577
11771 108.49423947841655	55115 234.7658407860905	88558 297.58696207999435
11881 109.0	55225 235.0	88668 297.7717246482614
11991 109.50342460398214	55335 235.23392612461325	88778 297.9563726453925
22002 148.3307115873176	55445 235.46761985462035	88888 298.14090628426015
22112 148.70104236352884	55555 235.70108188126758	88998 298.3253257770785
22222 149.07045314213008	55665 235.93431289238112	99009 314.65695606485485
22332 149.43895074578114	55775 236.16731357239087	99119 314.83170107217603
22442 149.80654191322887	55885 236.40008460235373	99229 315.0063491423625
22552 150.17323330074504	55995 236.63262665997686	99339 315.18090043655883
22662 150.53903148353254	66006 256.9163287920797	99449 315.35535511546334
22772 150.90394295710104	66116 257.13031715455105	99559 315.5297133393304
22882 151.2679741386127	66226 257.3441275801723	99669 315.70397526797154
22992 151.63113136819894	66336 257.557760512084	99779 315.8781410607578
33003 181.6672782864322	66446 257.77121639159014	99889 316.05221087662085
33113 181.96977771047588	66556 257.98449565816935	99999 316.226184874055
33223 182.27177510519834	66666 258.1975987494849	
33333 182.573272961844	66776 258.4105261013955	

Question 5 was given in a mathematics contest for 7-12th grade students. One produced this proof: The value of m must be between 100 and 316 to be a 5-digit number. Since a=e, the earliest possible forms of m are 101, 109, 111, ... since only $1^2=1$ and $9^2=81$ end in a one. The second one you try works, namely m=109. Very impressive analysis for a high-school mathematics student. PS: And no calculators were allowed on the contest.

This program was able to definitively find all of the solutions to the problem. In this case, the program can serve as a proof. Or, technically, the output itself can serve as a proof. Because there are a limited number of possible outcomes, the solution is complete and would satisfy even the most stringent of mathematicians.

Question 6 – Equal to the sum of the factorials of the digits

Which whole numbers are equal to the sum of the factorials of their digits? By hand calculations it is easy to see that 1 = 1! and 2 = 2! and that probably no 2-digit whole number would work.

```
import math
def sumFacDigits(n):
    strN = str(n)
    sum = 0
    for digit in strN:
        sum += math.factorial(int(digit))
    return sum
def main():
    stop=250000
    for i in range(1,stop):
        sfd = sumFacDigits(i)
        if i == sfd:
            print("It holds for i =",i)
main()
```

The resulting output is:

```
It holds for i = 1
It holds for i = 2
It holds for i = 145
It holds for i = 40585
```

The stop value in the main program was originally set to only 250, to get initial results. It was increased to 50000 and we hit the 40585 solution. There was a concern that raising the stop value further would result in an endlessly running program, but raising it to 250000 or a million still only took a few seconds, and no others were found. When programming in numbers this large, it can be an unknown on how long the program will require. A mathematician thinking along with a computer scientist would offer this thought: no eight-digit number will work, since 9!=362,880 and 8·9! is only 2,903,040. The largest eight-digit number is 99,999,999. Raising the limit to 10 million is sufficient to verify that there are exactly 4 solutions.

Question 8 – n!+1 is a perfect square

For which n is n! + 1 a perfect square? This is simply a brute-force algorithm that checks all values of n!+1, and reports which are perfect squares. And the results are all in the interpretation. This is a classic example where one should rely on an algorithm output for insight into a problem, but not necessarily for the direct solution itself. The program shows that only n = 4, 5, and 7 will work.

The brute-force algorithm is simple enough:

```
#Q8 -- For which n is n!+1 a perfect square?

import math

def isPerfectSquare(n):
    a = math.sqrt(n)
    b = int(math.sqrt(n))
    if a == b:
        return True
    return False

for i in range(1,100):
    factPlusOne = math.factorial(i) + 1
    print("i =",i," fact = ",factPlusOne-1," factPlusOne =",factPlusOne)
        if isPerfectSquare(factPlusOne):
            print("It holds for i =",i, )
```

Running from 1 to 10, it shows that it holds for 4, 5, and 7. These can easily be checked by hand.

```
i = 1    fact =    1    factPlusOne = 2
i = 2    fact =    2    factPlusOne = 3
i = 3    fact =    6    factPlusOne = 7
i = 4    fact =    24   factPlusOne = 25
It holds for i = 4
i = 5    fact =    120  factPlusOne = 121
It holds for i = 5
i = 6    fact =    720  factPlusOne = 721
i = 7    fact =    5040 factPlusOne = 5041
It holds for i = 7
i = 8    fact =    40320   factPlusOne = 40321
i = 9    fact =    362880  factPlusOne = 362881
i = 10   fact =    3628800 factPlusOne = 3628801
```

However, when the algorithm is run further, it starts to show its true weakness. After hitting i=30 and beyond, the program returns true for all numbers greater than 29. Is this the case? Unfortunately, none of these outputs are perfect squares.

```
i = 29   fact =   88417619937397019545436160000000   factPlusOne = 88417619937397019545436160000001
i = 30   fact =   2652528598121910586363084800000000   factPlusOne = 2652528598121910586363084800000001
It holds for i = 30
i = 31   fact =   82228386541779228177255628800000000   factPlusOne = 82228386541779228177255628800000001
It holds for i = 31
i = 32   fact =   2631308369336935301672180121600000000   factPlusOne = 2631308369336935301672180121600000001
It holds for i = 32
i = 33   fact =   86833176188118864955181944012800000000   factPlusOne = 86833176188118864955181944012800000001
It holds for i = 33
i = 34   fact =   295232799039604140847618609643520000000   factPlusOne = 295232799039604140847618609643520000001
It holds for i = 34
```

And so forth. In the immortal words of Tom, from Tom and Jerry: "DOOOOON'T YOU BELIEVE IT!" (YouTube link: https://www.youtube.com/watch?v=mG9ojyN216E) Note that these numbers are sufficiently high that hand-checking is not feasible. At this point, one can simply choose not to believe the results, and assume they are all false positive readings. The most likely culprit is a limitation in the `isPerfectSquare` function causing a false-positive result. That can be an exercise for the reader to explore.

Question 9 – For which n is F_n a perfect square?

For which n is F_n a perfect square? It is easy to see that $F_1 = 1$ and $F_2 = 1$ are perfect squares. But then what?

```
import math

def fibonacci(n):
    if n<0:
        print("Warning -- Incorrect input")
    elif n==1:
        return 1
    elif n==2:
        return 1
    else:
        return fibonacci(n-1)+fibonacci(n-2)


def isPerfectSquare(n):
    a = math.sqrt(n)
    b = int(math.sqrt(n))
    if a == b:
        return True
    return False


def main():
    for i in range(1,100):
        fib = fibonacci(i)
        if isPerfectSquare(fib):
            print(i, fib, math.sqrt(fib))

main()
```

The resulting output for up to F_{100} is the following. There is a published paper stating that 144 is the largest.

```
1 1 1.0
2 1 1.0
12 144 12.0
```

Question 10 – How many of 101, 10101, 1010101 ... are prime?

In fact all but 101 are composite. For ease of understanding this problem, you can expand in base 10 notation and then replace 10 by x. So 1010101 looks

97

like $10^6 + 10^4 + 10^2 + 1 = x^6 + x^4 + x^2 + 1$. Now this factors by grouping: $x^6 + x^4 + x^2 + 1 = x^4(x^2 + 1) + (x^2 + 1) = (x^2 + 1)(x^4 + 1)$ showing that 101 is a factor. Every number with an even number of ones factors by grouping. Those with an odd number of ones also factor:

$x^8 + x^6 + x^4 + x^2 + 1 = \frac{x^{10}-1}{x^2-1} = \frac{(x^5-1)(x^5+1)}{(x-1)(x+1)} = (x^4+x^3+x^2+x+1)(x^4-x^3+x^2-x+1)$. So 11,111 divides 101010101. Now let's see what a computer program does with this question.

```
#Q10 -- How many of 101, 10101, 1010101,... are prime?

def isPrime(n):
    "Brute-force algorithm to check for prime numbers"
    for i in range(2,n//2):
        if n % i == 0:
            return False
    return True

# isPrime2
# A method of checking for prime numbers, in which it stops at
# sqrt(n)), and only checks for primes of the form 6k +/- 1.
# The efficiency is exponentially greater than the other
function.
# Algorithm found on Stack Overflow at:
#     https://stackoverflow.com/questions/1801391/what-is-the-best-algorithm-for-checking-if-a-number-is-prime
def isPrime2(n):
    """Returns True if n is prime."""
    if n == 2:
        return True
    if n == 3:
        return True
    if n % 2 == 0:
        return False
    if n % 3 == 0:
        return False
    i = 5
    w = 2
    while i * i <= n:
        if n % i == 0:
            return False
        i += w
        w = 6 - w
    return True

def prime10101(start, end):
```

```
        numString = "101"
        for i in range(start-1):
            numString += "01"
        for i in range(start, end+1):
            num = int(numString)
            print("Doing iteration",i, num)
            if isPrime2(num):
                print(num,"is prime!")
            numString += "01"

prime10101(1,17)
print()
#prime10101(18,18)   # could be prime, hangs program, even with
the efficient algorithm
print()
prime10101(19,100)
```

The following is the output of the program above. Note that iteration 18 has been skipped. For some reason iteration 18 would unexpectedly run the program indefinitely. According to the proof discussed above, iteration 18 should be composite. The results below show that only 101 is prime, and all other formats are composite.

```
Doing iteration 1 101
101 is prime!
Doing iteration 2 10101
Doing iteration 3 1010101
Doing iteration 4 101010101
Doing iteration 5 10101010101
Doing iteration 6 1010101010101
Doing iteration 7 101010101010101
Doing iteration 8 10101010101010101
Doing iteration 9 1010101010101010101
Doing iteration 10 101010101010101010101
Doing iteration 11 10101010101010101010101
Doing iteration 12 1010101010101010101010101
Doing iteration 13 101010101010101010101010101
Doing iteration 14 10101010101010101010101010101
Doing iteration 15 1010101010101010101010101010101
Doing iteration 16 101010101010101010101010101010101
Doing iteration 17 10101010101010101010101010101010101

Doing iteration 19 1010101010101010101010101010101010101010101
Doing iteration 20 101010101010101010101010101010101010101010101
Doing iteration 21 10101010101010101010101010101010101010101010101
```

It is easy to see that this question will challenge the program.

Question 12 – Inequality by mathematical induction

What happens when you try to prove $(n - 1) / (n + 1) < 9/10$. This is a trap problem. When a student sees the variable **n** in a problem, mathematical induction springs to mind. One checks a few initial cases for **n** to see if it is feasible; In fact it falls apart at **n** = 20. It is tedious to substitute by hand n=1, 2, ..., 20.

```
def main():
    for i in range(1,30):
        left = (i-1)/(i+1)
        right = 9/10
        if left < right:
            inequality = "   <   "
        elif left > right:
            inequality = "   >   "
        else:
            inequality = "   ==  "
        print("i = %3d    %.3f %6s %.1f" % (i, left, inequality, right))

main()
```

The resulting output is below. Note that from n=20 on, the expression is no longer less than 0.9.

```
i =   1    0.000   <    0.9
i =   2    0.333   <    0.9
i =   3    0.500   <    0.9
i =   4    0.600   <    0.9
i =   5    0.667   <    0.9
i =   6    0.714   <    0.9
i =   7    0.750   <    0.9
i =   8    0.778   <    0.9
i =   9    0.800   <    0.9
i =  10    0.818   <    0.9
i =  11    0.833   <    0.9
i =  12    0.846   <    0.9
i =  13    0.857   <    0.9
i =  14    0.867   <    0.9
i =  15    0.875   <    0.9
i =  16    0.882   <    0.9
i =  17    0.889   <    0.9
i =  18    0.895   <    0.9
```

```
i = 19    0.900    ==    0.9
i = 20    0.905    >     0.9
i = 21    0.909    >     0.9
i = 22    0.913    >     0.9
i = 23    0.917    >     0.9
i = 24    0.920    >     0.9
i = 25    0.923    >     0.9
i = 26    0.926    >     0.9
i = 27    0.929    >     0.9
i = 28    0.931    >     0.9
i = 29    0.933    >     0.9
```

Of course, using algebraic steps on the inequality also shows n < 19.

Question 13 – Mersenne Primes

Here is a theorem that is not difficult to prove, using the contrapositive. If 2^n-1 is prime, then n is prime. However, is the converse true? The converse would be: If n is prime, then 2^n-1 is prime. It would be helpful for ease of computation to have a program generate the data and continue the table. A hand-held calculator is also sufficient for ease of computation.

n	2	3	5	7	11	13
2^n-1	3	7	31	127	2047	8191

Since $2047 = 23 \cdot 89$, it is not prime and the converse is false. An iPhone can show that 8191 is prime. In fact, it is a Mersenne prime – a prime that is one less than a power of 2. The first few Mersenne primes are: 3, 7, 31, 127, 8191. As of October 2020, there are only 51 known Mersenne primes, although the suspicion is that there are infinitely many.

Chapter 6 – Induction

Constructing a proof by mathematical induction is very much like writing your first (short) English composition paper. There you learned to form an outline first, carefully identifying the beginning, main body, and ending. Paying close attention to the rules of grammar and writing, you forged ahead and produced a 300-word paper that was typically shredded to pieces by a seemingly overly critical English professor. After a semester of this, however, most could write a coherent scholarly paper.

Treat your experience of learning to write a proof by mathematical induction the same way. Each proof is, after all, a somewhat complicated thought that you wish conveyed in writing. Your outline should contain the following steps:

1. Clearly describe what the proposition p(n) is.
2. State that you are going to prove p(n) true for n ∈ X = {a, a + 1, ...}, clearly indicating what the integer a is.
3. Show that p(a) is true.
4. Assume that p(k) is true for (some single) k ∈ X. Say what this means for the proposition at hand. (This is often called the INDUCTIVE HYPOTHESIS.)
5. Say that you need to show p(k + 1) true and translate to what this means in terms of the problem at hand.
6. Show p(k + 1) to be true, using p(k).
7. State that since p(k) implies p(k + 1), the proposition is true for all n ∈ X by the principle of mathematical induction.

The following model proof illustrates the features of this outline. After some practice, you will find yourself compressing these steps into fewer, but essential steps.

As you read each of the following proofs, it may help to think of the process as being analogous to climbing a ladder.

STEP 1: The BASIS needs to be established – you need to first get on the ladder, on some rung not necessarily the first rung.

STEP 2: The INDUCTIVE HYPOTHESIS, assume p(k) is true for some k-- imagine that you can get onto rung k.

STEP 3: Show that p(k + 1) is true – show that you can get to rung k + 1 if you are standing on rung k.

Note: It may be helpful to think of k as being equal to 801 in every problem. That way, you can think of k as being fixed, but arbitrary. Proof by mathematical induction is a very algorithmic process and it is important that each step of the process is done carefully and precisely. You must establish a basis, otherwise you have no foundation upon which to build. Once you have a foundation in place, you must make sure that each step follows from the previous one, otherwise your structure collapses. A second analogy often used to represent proof by mathematical induction involves a row of dominoes. If you can knock down the first domino and the dominoes are spaced so that each domino is knocked down by the one which precedes it, then the entire row of dominoes will fall.

What follows is a model to start you off. It includes complete details: Stating p(n), listing the three steps, and then executing the steps. As you read through the many examples that follow this model you will often notice missing steps, missing format items, missing reasons…. This is typical as you gain ever more experience with proofs by mathematical induction.

A Model Proof

$$p(n): 1 \cdot 1! + 2 \cdot 2! + 3 \cdot 3! + \cdots + n \cdot n! = (n+1)! - 1, n \geq 1$$

<u>STEP 1:</u> p(1) is true since LHS = $1 \cdot 1!$ and RHS = $(1 + 1)! - 1 = 2! - 1 = 1$

<u>STEP 2:</u> Assume p(k) is true for some, one, fixed, but arbitrary value k ∈ {1, 2, 3, …}; i.e. assume $1 \cdot 1! + 2 \cdot 2! + 3 \cdot 3! + \cdots + k \cdot k! = (k+1)! - 1$ for that one k.

<u>STEP 3:</u> Prove that p(k + 1) is true using the <u>assumption</u> that p(k) is true; i.e., you need to prove that

$$1 \cdot 1! + 2 \cdot 2! + \cdots + k \cdot k! + (k+1) \cdot (k+1)! = (k+2)! - 1.$$

(Always keep your eye on step 3!)

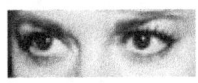

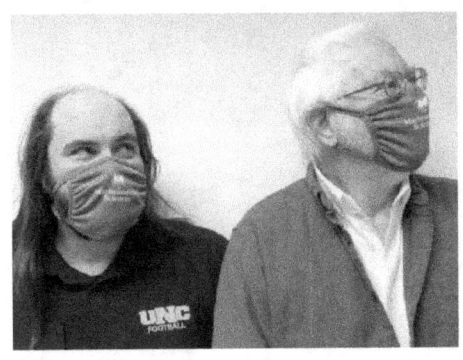

NOW START: Starting with the LHS of what is to be proved, we have

$$1 \cdot 1! + 2 \cdot 2! + \cdots + k \cdot k! + (k+1) \cdot (k+1)!$$

$$= (k+1)! - 1 + (k+1)(k+1)! \qquad \text{(Using Step 2)}$$

$$= (k+1)! \, [1 + k + 1] - 1 \qquad \text{(Factoring)}$$

$$= (k+1)! \, (k+2) - 1$$

$$= (k+2)! - 1$$

Since $p(k) \Rightarrow p(k+1)$ and k is arbitrary, $p(n)$ is true for all $n \in \{1, 2, 3, \ldots\}$ by the principle of mathematical induction.

SAMPLE PROOFS

1. Prove that n straight lines in a plane, passing through one point P, divide the plane into 2n regions.

    Let p(n) denote that statement. p(1) is certainly true since 1 line splits the plane into 2 regions. Now assume that k lines divide the plane into 2k regions. At this stage you might draw the situation representing n = 3 or n = 4 lines. Now draw a (k + 1)st line through the point P, splitting one region into two and as that line emerges from P splits the corresponding region in two. Now the total number of regions is

    $$2k + 1 + 1 = 2k + 2 = 2(k + 1).$$

    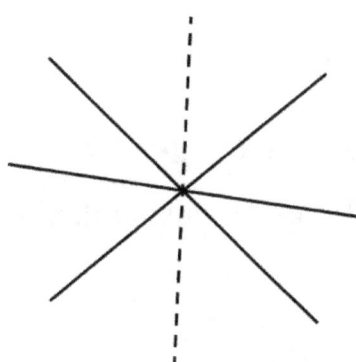

2. Prove that $\ln(a_1 a_2 a_3 \ldots a_n) = \ln(a_1) + \ln(a_2) + \ln(a_3) + \cdots + \ln(a_n)$ using $\ln(xy) = \ln(x) + \ln(y)$.

    First the statement certainly holds for n = 1. Now assume that

    $$\ln(a_1 a_2 a_3 \ldots a_k) = \ln(a_1) + \ln(a_2) + \ln(a_3) + \cdots + \ln(a_k).$$

    Then, $\ln(a_1 a_2 a_3 \ldots a_k a_{k+1}) = \ln(a_1 a_2 \ldots a_k)(a_{k+1})$

    $$= \ln(a_1 a_2 a_3 \ldots a_k) + \ln(a_{k+1})$$

    $$= \ln(a_1) + \ln(a_2) + \cdots + \ln(a_k) + \ln(a_{k+1})$$

    Since k is arbitrary, the original statement is true for all n.

3. Prove that the sum of the cubes of three consecutive positive integers is divisible by 9.

Strategic choice of the form of these three integers might be helpful. So, let p(n) be the proposition that $(n-1)^3 + n^3 + (n+1)^3$ is a multiple of 9. We prove p(n) true for $n \geq 1$. p(1) holds since $0^3 + 1^3 + 2^3 = 9$. Now assume that $(k-1)^3 + k^3 + (k+1)^3 = 3k(k^2 + 2) = 3k^3 + 6k$ is divisible by 9.

Then, $k^3 + (k+1)^3 + (k+2)^3$

$$= k^3 + k^3 + 3k^2 + 3k + 1 + k^3 + 6k^2 + 12k + 8$$

$$= 3k^3 + 9k^2 + 15k + 9$$

$$= 3k^3 + 9k^2 + 6k + 9k + 9 = (3k^3 + 6k) + 9(k^2 + k) + 9$$

is the sum of multiples of 9 using the inductive hypothesis and we are done.

Here is an alternate algebra proof using the fact that the product of three consecutive integers is always divisible by 3:

$$(n-1)^3 + n^3 + (n+1)^3 = 3n(n^2 + 2) = 3n(n^2 - 1 + 3)$$

$$= 3n(n^2 - 1) + 9n$$

$$= 3(n-1)(n)(n+1) + 9n$$

As you read through this proof, feel free to fill in the missing algebra steps:

$$(n-1)^3 = ?$$

4. Here we prove the power rule for differentiation
$$(x^n)' = nx^{n-1} \text{ for } n \geq 1$$
using the product rule.

The base case of $n = 1$ is a little tricky as you cannot use the power rule. You could say that the slope of the line $y = x$ is 1 or use the definition of the derivative to conclude that if $f(x) = x$, then $f'(x) = 1$.

Now assume that $(x^k)' = kx^{k-1}$.

Then, $(x^{k+1})' = (x \cdot x^k)'$

$= x \cdot kx^{k-1} + x^k \cdot 1$

$= kx^k + x^k = (k+1)x^k$

using the product rule on x^{k+1} and the inductive hypothesis.

5. Given an unlimited supply of 5¢ and 9¢ postage stamps, lets prove here by induction that n¢ of postage can be made using these denominations for n ∈ {32, 33, ...}. First, 32¢ can be made using three 9¢ stamps and one 5¢ stamp. Now, assume that k¢ can be so assembled. In that pile or collection of k¢, you either have

   a) At least one 9¢ stamp or
   b) No 9¢ stamps.

In the first case, pull out that one 9¢ stamp and replace it with two 5¢ stamps, thereby increasing the k¢ pile to $(k + 1)$¢. In the second case, if there are no 9¢ stamps, there must be at least seven 5¢ stamps (remember that n ≥ 32). So, replace these with four 9¢ stamps, increasing the k¢ pile by 1¢ again. You now have (k + 1)¢, completing the induction.

How about a non-induction proof, like you might see in the 5th grade?

First get five in a row:

$$32 = 3 \cdot 9 + 1 \cdot 5$$

$$33 = 2 \cdot 9 + 3 \cdot 5$$

$$34 = 1 \cdot 9 + 5 \cdot 5$$

$$35 = \phantom{0 \cdot 9 + {}} 7 \cdot 5$$

$$36 = 4 \cdot 9$$

Now, if you just add a 5¢ stamp to each you can make

$$37, 38, 39, 40, 41.$$

Now, add another 5¢ stamp to each, or start with nine in a row and add a 9¢ stamp to each.

Why start with 32¢? $(9 - 1)(5 - 1) = 8 \cdot 4 = 32$.

In general, if you have postage values of a¢ and b¢, you can make any amount from $(a - 1)(b - 1)$ on. Of course, don't set a = 4¢ and b = 6¢, or a = 3¢ and b = 9¢. They need to be relatively prime. See Chapter 5 for a computer program to solve this problem for any stamp values.

6. Prove that $n! > 3^n$. Use a data table here to first determine initial n.

n	1	2	3	4	5	6	7
n!	1	2	6	24	120	720	5040
3^n	3	9	27	81	243	729	2187

From this table, it looks like n! catches up to 3^n when n = 7.

So, it looks like we want to prove $n! > 3^n$ for $n \geq 7$.

We have $(k+1)! = (k+1)k! > (k+1)3^k > 3 \cdot 3^k = 3^{k+1}$, since $k \geq 7$.

Interpretation: A factorial function grows faster than an exponential function.

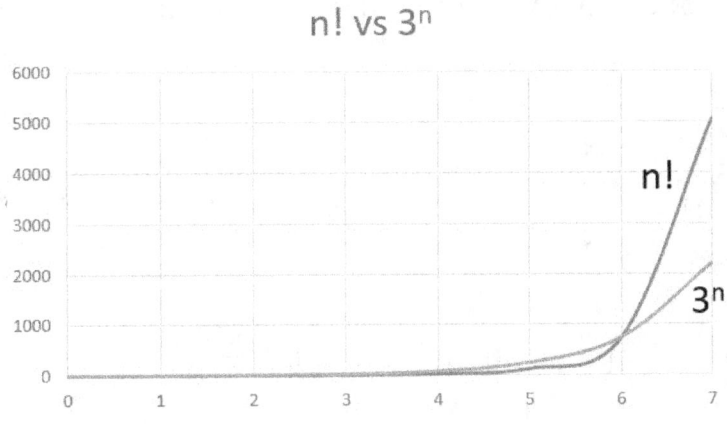
n! vs 3^n

7. Let p(n) be the proposition that $10^{n+1} + 10^n + 1$ is divisible by 3 for $n \geq 0$. p(0) is true, since 12 has 3 as a factor. Now assume that $10^{k+1} + 10^k + 1$ is divisible by 3. Then,

$$10^{k+2} + 10^{k+1} + 1 = 10 \cdot 10^{k+1} + 10 \cdot 10^k + 10 - 9$$
$$= 10(10^{k+1} + 10^k + 1) - 9$$

Now each of these two terms is divisible by 3.

8. Let p(n) be the proposition that $2^n > n^2$. We prove that p(n) holds for n = 1 and n > 4. For n = 2, 3, 4, this inequality is not true. Check this. p(5) is true since 32 > 25. Now assume that $2^k > k^2$ (for some arbitrary k > 4).

$$\begin{aligned}\text{Then, } 2^{k+1} = 2 \cdot 2^k &> 2k^2 \\ &= k^2 + k^2 \geq k^2 + 5k \quad \text{(since } k \geq 5\text{)} \\ &= k^2 + 2k + 3k > k^2 + 2k + 1 \quad \text{(since } 3k > 1\text{)} \\ &= (k+1)^2 \end{aligned}$$

Interpretation: An exponential grows faster than a quadratic.

9. Conjecture and prove by induction a formula for D(n), the number of diagonals of a convex n-gon. First, drawing lots of pictures will yield this table:

n	3	4	5	6	7	...
D(n)	0	2	5	9	14	...

Let p(n) be the proposition that $D(n) = \frac{n(n-3)}{2}$

1. p(3) is true since $D(3) = 0$ and $\frac{3(3-3)}{2} = 0$

2. Assume that a k-gon has $\frac{k(k-3)}{2}$ diagonals.

3. Prove (or show) that $(k+1)$ − gon has $\frac{(k+1)(k-2)}{2}$ diagonals.

Draw a k-gon:

Break a side:

Count the old and new diagonals:

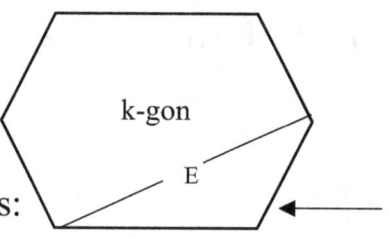

k-gon

E

(k + 1)st vertex

The k-gon has $\frac{k(k-3)}{2}$ diagonals, by the inductive hypothesis. There are $(k+1) - 3$ new diagonals from the new vertex. Also edge E is now a diagonal.

$$\text{Total} = \frac{k(k-3)}{2} + (k+1) - 3 + 1$$

$$= \frac{k(k-3)}{2} + k - 1 = \frac{k(k-3)}{2} + \frac{2k-2}{2}$$

$$= \frac{k^2 - 3k + 2k - 2}{2} = \frac{k^2 - k - 2}{2}$$

$$= \frac{(k-2)(k+1)}{2}$$

Since $p(k) \Rightarrow p(k+1)$, $p(n)$ is true for all $n \in \{3, 4, 5, ...\}$ by the principle of mathematical induction.

Here is a simpler non-induction proof. Each of the n vertices can be connected to any of the $n - 3$ non-adjacent vertices for a total of $n(n-3)$ choices. However, connecting vertex A to vertex B is the same as connecting B to A. So, these are duplicate diagonals. Therefore, the answer is $\frac{n(n-3)}{2}$.

10. Conjecture a formula in terms of Fibonacci numbers for

$$S_n = F_1^2 + F_2^2 + F_3^2 + \cdots + F_n^2$$

The data table

n	1	2	3	4	5	6
F_n	1	1	2	3	5	8
S_n	1	2	6	15	40	104

will help you decide on a nice formula, that could easily be proved by induction. Alternatively, you could use $F_n^2 = F_n(F_{n+1} - F_{n-1})$ and telescope. Or, you could try the very familiar geometry "proof": Assemble the squares into a rectangle.

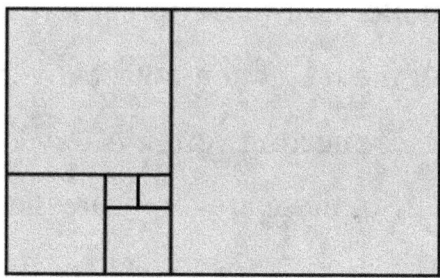

11. Let p(n) be the proposition that $n^3 + 2n$ is a multiple of 3 for $n \geq 0$.

p(0) is certainly true. Now, assume that $k^3 + 2k$ is a multiple of 3 and using that, conclude $(k + 1)^3 + 2(k + 1)$ is also a multiple of 3.

$$(k + 1)^3 + 2(k + 1) = k^3 + 3k^2 + 3k + 1 + 2k + 2$$

$$= \left(k^3 + 2k\right) + (3k^2 + 3k + 3).$$

These two terms are each divisible by 3.

12. Prove that $F_0 + F_2 + F_4 + \cdots + F_{2n} = F_{2n+1} - 1, n \geq 0$. Here, F_n stands for the nth Fibonacci Number.

So, let p(n) be the statement $F_0 + F_2 + F_4 + \cdots + F_{2n} = F_{2n+1} - 1$. We will show that p(n) is true for all $n \geq 0$. The base case is easy, since $F_0 = 0$ and $F_1 = 1$, so $F_0 = F_1 - 1$. Now, consider the inductive case: assume p(k) is true. That is, assume $F_0 + F_2 + F_4 + \cdots + F_{2k} = F_{2k+1} - 1$. To establish p(k + 1), we work from left to right:

$$F_0 + F_2 + \cdots + F_{2k} + F_{2k+2} = F_{2k+1} - 1 + F_{2k+2} \quad \text{(by the ind. hypothesis)}$$

$$= F_{2k+1} + F_{2k+2} - 1$$

$$= F_{2k+3} - 1 \quad \text{(by recursive def.)}$$

Therefore, $F_0 + F_2 + F_4 + \cdots + F_{2k+2} = F_{2k+3} - 1$, which is to say that p(k + 1) holds. Thus, by the principle of mathematical induction, p(n) is true for all $n \geq 0$.

13. Prove that $n^3 - 4n + 6$ is divisible by 3 for all natural numbers n, by mathematical induction.

Since this statement is true for n = 0, we now assume that $k^3 - 4k + 6$ is a multiple of 3. Then,

$$(k + 1)^3 - 4(k + 1) + 6 = k^3 + 3k^2 + 3k + 1 - 4k - 4 + 6$$
$$= (k^3 - 4k + 6) + 3(k^2 + k - 1),$$

which is the sum of two multiples of 3.

Here is a non-induction proof, which uses a little bit of clever algebra. Recalling that the product of three consecutive integers is always a multiple of three,

$$n^3 - 4n + 6 = n(n^2 - 1) - 3n + 6$$
$$= (n-1)(n)(n+1) - 3(n-2),$$

which is a multiple of 3.

14. Prove that $1 \cdot 2 + 2 \cdot 3 + \cdots + n(n+1) = \frac{n(n+1)(n+2)}{3}$.

A proof by mathematical induction is left for you in the exercises. For now, here is a non-induction proof:

Let $S = 1 \cdot 2 + 2 \cdot 3 + \cdots + n(n+1)$. Now divide both sides by 2.

$$\frac{S}{2} = \frac{1 \cdot 2}{2} + \frac{2 \cdot 3}{2} + \cdots + \frac{n(n+1)}{2} = \binom{2}{2} + \binom{3}{2} + \cdots + \binom{n+1}{2}$$
$$= \binom{n+2}{3} = \frac{(n+2)(n+1)(n)}{6}.$$

Now, solve for S.

15. Prove that $(1 + \alpha)^n > 1 + \alpha n$ where $\alpha > -1, \alpha \neq 0, n \geq 2$.

Here $\alpha \in \mathbb{R}$. Assume $(1+\alpha)^k > 1 + \alpha k$. Then

$$(1+\alpha)^{k+1} = (1+\alpha)(1+\alpha)^k > (1+\alpha)(1+k\alpha) \quad \text{(by hypothesis)}$$

$$= 1 + k\alpha + \alpha + k\alpha^2 > 1 + (k+1)\alpha \quad \text{(since } k\alpha^2 > 0\text{)}$$

Alternatively, here is a simpler, non-induction proof:

$$(1+\alpha)^n = \binom{n}{0} + \binom{n}{1}\alpha + \binom{n}{2}\alpha^2 + \cdots + \binom{n}{n}\alpha^n > \binom{n}{0} + \binom{n}{1}\alpha = 1 + \alpha n.$$

16. We leave the induction proof of $1^2 + 2^2 + 3^2 + \cdots + n^2 = \frac{n(n+1)(2n+1)}{6}$ to you in the exercises. Instead, here is a fun non-induction proof:

Let's use $1 \cdot 2 + 2 \cdot 3 + \cdots + n(n+1) = \frac{n(n+1)(n+2)}{3}$.

$$1 \cdot 2 + 2 \cdot 3 + 3 \cdot 4 + \cdots + n(n+1)$$
$$= 1(1+1) + 2(1+2) + 3(1+3) + \cdots + n(1+n)$$
$$= (1 + 2 + 3 + \cdots + n) + 1^2 + 2^2 + 3^2 + \cdots + n^2. \text{ Then,}$$

$$1^2 + 2^2 + \cdots + n^2 = \frac{n(n+1)(n+2)}{3} - \frac{n(n+1)}{2} = \frac{n(n+1)(2n+1)}{6}.$$

Then another proof, using $n^2 = \binom{n}{2} + \binom{n+1}{2}$; (This little result is often read as "the sum of two consecutive triangular numbers is a perfect square" – see picture 6.)

$$1^2 + 2^2 + \cdots + n^2 = \binom{2}{2} + \binom{2}{2} + \binom{3}{2} + \binom{3}{2} + \cdots + \binom{n}{2} + \binom{n+1}{2}$$

$$= \binom{2}{2} + \binom{3}{2} + \cdots + \binom{n}{2} + \binom{2}{2} + \binom{3}{2} + \cdots + \binom{n+1}{2}$$

$$= \binom{n+1}{3} + \binom{n+2}{3}$$

$$= \frac{n(n+1)(2n+1)}{6}.$$

And next, a "picture proof" illustrated with a 5×5 square. Look along diagonal D. Each choice of 2 of the 4 lattice points determines exactly one square. One point fixes the lower left corner and the other one the upper right corner. So diagonal D determines $\binom{4}{2} = 6$ squares. Repeat for each diagonal and get

$\binom{2}{2} + \binom{3}{2} + \binom{4}{2} + \binom{5}{2} + \binom{6}{2}$ PLUS $\binom{5}{2} + \binom{4}{2} + \binom{3}{2} + \binom{2}{2}$. Now apply the Hockey-Stick theorem twice to get $\binom{7}{3} + \binom{6}{3}$. A similar general proof will give $\binom{n+1}{3} + \binom{n+2}{3}$ as the total number of squares of all sizes in an n×n square.

5×5 square

D

As a bonus we get that the number of squares (again all sides are parallel to sides of the original square) in a subdivided n×(n+1) rectangle is $2\binom{n+2}{3}$. The 2 comes from the fact that the diagonal D is now duplicated. See diagram.

4×5 rectangle

D   D

And finally, this one will pique your interest.

Here is a new (KAJ) proof for $1^2 + 2^2 + 3^2 + \ldots + n^2 = \binom{n+1}{3} + \binom{n+2}{3}$ for $n = 8$, the checkerboard problem. Counting all sizes from size 1×1 to 8×8 there are a total of $1^2 + 2^2 + \ldots + 8^2$ squares. Now we introduce a new tool.

A 3-digit number like 687 is a <u>peak number</u>, denoted as mpn, if its digits strictly increase to the middle and then strictly descend. It is not hard to see that there are $2\binom{9}{3}$ peak numbers; just choose 3 digits from $\{1, 2, 3, \ldots, 9\}$ and then swap m and n. There is a nice correspondence between certain peak numbers and squares in the 8 by 8 array.

First embed the checkerboard into the first quadrant so that its lower left corner (LLC) sits at the lattice point (1,1). The peak number mpn corresponds with the square whose LLC is (m,n) and has side lengths $s = p - \max\{m,n\}$. 687 and its square is shown. The remaining missing squares are those where LLC is on the diagonal $y = x$ matching with peaks like mpm which were not included when choosing 3 from $\{1, 2, 3, \ldots, 9\}$. There are $\binom{9}{2}$ of this type. The square 363 is shown. Then the total number of squares in our checkerboard is $2\binom{9}{3} + \binom{9}{2} = \binom{10}{3} + \binom{9}{3}$.

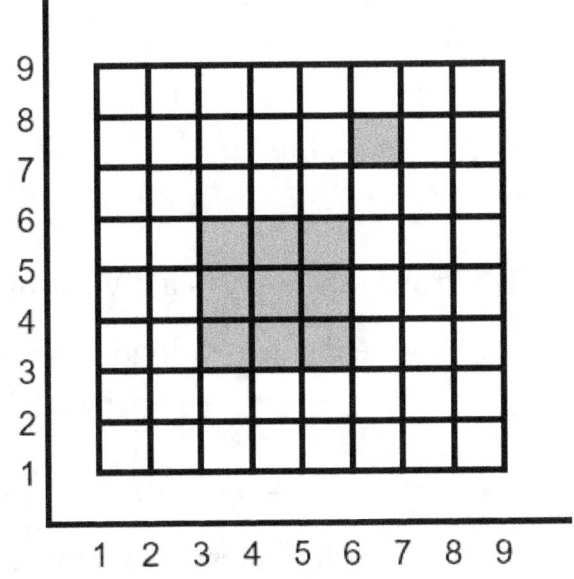

17. Let $T_n = \left(1 - \frac{1}{4}\right)\left(1 - \frac{1}{9}\right)\left(1 - \frac{1}{16}\right) \cdots \left(1 - \frac{1}{n^2}\right)$. Conjecture and prove a nice formula for T_n.

n	2	3	4	5	6	...
T_n	$\frac{3}{4}$	$\frac{2}{3}$	$\frac{5}{8}$	$\frac{3}{5}$	$\frac{7}{12}$	...

This is a little tricky, but manageable if you "un-reduce" T_n for n odd; e.g. $\frac{2}{3} = \frac{4}{6}$. Then, T_n looks like $\frac{n+1}{2n}$.

So, p(n): $\left(1 - \frac{1}{4}\right)\left(1 - \frac{1}{9}\right) \cdots \left(1 - \frac{1}{n^2}\right) = \frac{n+1}{2n}$, $n \geq 2$. p(2) is true—look at the data table. Then,

$$\left(1 - \frac{1}{4}\right)\left(1 - \frac{1}{9}\right) \cdots \left(1 - \frac{1}{k^2}\right)\left(1 - \frac{1}{(k+1)^2}\right) = \frac{k+1}{2k}\left(1 - \frac{1}{(k+1)^2}\right)$$

$$= \frac{k+2}{2(k+1)} \qquad \text{(after a bunch of algebra steps)}$$

This does it!

18. Prove that $x - y$ is a divisor of $x^n - y^n$ for $n \geq 0$.

Assume that $x - y$ divides $x^k - y^k$. Then,

$$x^{k+1} - y^{k+1} = x \cdot x^k - y \cdot x^k + y \cdot x^k - y \cdot y^k$$
$$= (x - y)x^k + y(x^k - y^k).$$

Here, each of these two terms is divisible by $x - y$. Alternatively, you could also add in and subtract out xy^k. Try it!

In the 10th grade, you may have done this by long division to arrive at the following:

$$\frac{x^n - y^n}{x - y} = x^{n-1} + x^{n-2}y^1 + \cdots + x^1 y^{n-2} + y^{n-1}.$$

Notice that this quotient is a geometric sum with ratio $r = \frac{y}{x}$. A shift and subtract technique will demonstrate that $x - y$ divides $x^n - y^n$:

$$S = x^{n-1} + x^{n-2}y^1 + \cdots + x^1 y^{n-2} + y^{n-1}$$

$$\frac{y}{x}S = x^{n-2}y^1 + \cdots + y^{n-1} + \frac{y^n}{x}$$

$$S - \frac{y}{x}S = x^{n-1} - \frac{y^n}{x}$$

$$(x - y)S = x^n - y^n$$

Finally, $S = \dfrac{x^n - y^n}{x - y}$.

19. Prove that 5 is a divisor of $9^n - 4^n$, $n \geq 0$.

Assume that $9^k - 4^k$ is a multiple of 5. Then,

$$\begin{aligned} 9^{k+1} - 4^{k+1} &= 9 \cdot 9^k - 4 \cdot 4^k \\ &= 9 \cdot 9^k - 9 \cdot 4^k + 5 \cdot 4^k \\ &= 9(9^k - 4^k) + 5 \cdot 4^k \end{aligned}$$

Each term is divisible by 5, so we are done.

Here is an easier, non-inductive proof: using the fact that $x - y$ is a divisor of $x^n - y^n$; just let $x = 9$ and let $y = 4$. This technique yields many similar results; 7 divides $12^n - 5^n$, 3 divides $17^n - 14^n$,...

20. Show that $S = 1^n + 8^n - 3^n - 6^n$ is a multiple of 10 for $n \geq 0$.

First, observe that $8^n - 3^n$ is a multiple of 5, as is $1^n - 6^n$. So, S is a multiple of 5. Now, rearrange your thoughts: $8^n - 6^n$ is a multiple of 2, as is $1^n - 3^n$. So now, S is divisible by both 2 and 5; hence, it is also divisible by 10.

Mathematical induction did the heavy lifting here: $x - y$ is a divisor of $x^n - y^n$.

Many related divisibility problems can be easily resolved using this general result.

21. Here is a pretty useless, but fun fact: 133 is always a divisor of $11^{n+2} + 12^{2n+1}$, $n \geq 0$.

Assume $11^{k+2} + 12^{2k+1}$ is a multiple of 133. Then,

$$11^{k+3} + 12^{2k+3} = 11 \cdot 11^{k+2} + 144 \cdot 12^{2k+1}$$
$$= 11 \cdot 11^{k+2} + 11 \cdot 12^{2k+1} + 133 \cdot 12^{2k+1}$$

You can see where this is going!

22. Prove that $\binom{n}{0} + \frac{1}{2}\binom{n}{1} + \frac{1}{3}\binom{n}{2} + \cdots + \frac{1}{n+1}\binom{n}{n} = \frac{2^{n+1}-1}{n+1}$.

We leave the proof by mathematical induction for you in the exercises. However, here is a calculus-type resolution of this identity:

The appearance of terms such as $\frac{1}{2}, \frac{1}{3}, \ldots, \frac{1}{n+1}$ might remind one of integration in calculus. So, let's start with this version of the binomial theorem:

$$(1+x)^n = \binom{n}{0} + \binom{n}{1}x + \binom{n}{2}x^2 + \cdots + \binom{n}{n}x^n$$

and integrate:

$$\frac{(1+x)^{n+1}}{n+1} = \binom{n}{0}x + \binom{n}{1}\frac{x^2}{2} + \cdots + \binom{n}{n}\frac{x^{n+1}}{n+1} + C.$$

When $x = 0$, $C = \frac{1}{n+1}$. Now, let $x = 1$.

23. Prove that $\binom{n}{1} + 2\binom{n}{2} + 3\binom{n}{3} + \cdots + n\binom{n}{n} = n2^{n-1}$.

    This identity can be handled in a variety of ways. The easiest is to differentiate

    $$(1 + x)^n = \binom{n}{0} + \binom{n}{1}x + \cdots + \binom{n}{n}x^n$$

    and then let $x = 1$:

    $$n(1 + x)^{n-1} = \binom{n}{1} + 2\binom{n}{2}x + \cdots + n\binom{n}{n}x^{n-1}$$

    $$n2^{n-1} = \binom{n}{1} + 2\binom{n}{2} + \cdots + n\binom{n}{n}$$

    A second way is to make use of the identity $k\binom{n}{k} = n\binom{n-1}{k-1}$:

    $$1\binom{n}{1} + 2\binom{n}{2} + \cdots + n\binom{n}{n}$$

    $$= n\binom{n-1}{0} + n\binom{n-1}{1} + \cdots + n\binom{n-1}{n-1}$$

    $$= n2^{n-1}.$$

    The next way is a surprise! Using symmetry, $\binom{n}{k} = \binom{n}{n-k}$, you can reverse and add these terms as you would do with an arithmetic sum. Try it!

    Finally, a fourth way: mathematical induction. This method is a little tedious, but straightforward. Just use the recursion on each term, such as $\binom{k+1}{3} = \binom{k}{2} + \binom{k}{3}$, and make two sums. Complete this in the exercises.

24. A hockey stick result. Show that $\binom{6}{0} + \binom{7}{1} + \binom{8}{2} + \cdots + \binom{n+6}{n} = \binom{n+7}{n}$ for $n \geq 0$.

Assume that $\binom{6}{0} + \binom{7}{1} + \binom{8}{2} + \cdots + \binom{k+6}{k} = \binom{k+7}{k}$. Then,

$$\binom{6}{0} + \binom{7}{1} + \cdots + \binom{k+6}{k} + \binom{k+7}{k+1} = \binom{k+7}{k} + \binom{k+7}{k+1}$$

(using the I.H.)

$$= \binom{k+8}{k+1} \text{ (by recursion for } \binom{n}{k}\text{)}.$$

Using the symmetry formula $\binom{n}{k} = \binom{n}{n-k}$, we get an alternate version for free:

$$\binom{6}{6} + \binom{7}{6} + \binom{8}{6} + \cdots + \binom{n+6}{6} = \binom{n+7}{7}.$$

25. Let p(n) be the proposition that $\frac{1}{1\cdot 3} + \frac{1}{3\cdot 5} + \cdots + \frac{1}{(2n-1)(2n+1)} = \frac{n}{2n+1}$.

Let's show that p(n) is true for $n \geq 1$. First, p(1) holds since $\frac{1}{1\cdot 3} = \frac{1}{1+2}$.

Now, assume p(k) is true. Then,

$$\frac{1}{1\cdot 3} + \frac{1}{3\cdot 5} + \cdots + \frac{1}{(2k-1)(2k+1)} + \frac{1}{(2k+1)(2k+3)} = \frac{k}{2k+1} + \frac{1}{(2k+1)(2k+3)}$$

$$= \frac{k(2k+3)}{(2k+1)(2k+3)} + \frac{1}{(2k+1)(2k+3)}$$

$$= \frac{2k^2+3k+1}{(2k+1)(2k+3)}$$

$$= \frac{k+1}{2k+3},$$

showing that p(k + 1) is true.

26. This rule of exponents will bring back old memories from the eighth grade. If a and b are constants, then $(ab)^n = a^n b^n$. So, let p(n) be the proposition that

$(ab)^n = a^n b^n$. We will show by induction that p(n) holds for $n \geq 0$.

p(0) is certainly true, since $(ab)^0 = 1 = a^0 b^0 = 1 \cdot 1$. Now, assume that p(k) is true: that is, $(ab)^k = a^k b^k$ for some fixed arbitrary $k \geq 0$. Now, proceed to show that p(k + 1) is true:

$$(ab)^{k+1} = (ab)(ab)^k$$
$$= a \cdot b \cdot a^k \cdot b^k = a \cdot a^k \cdot b \cdot b^k$$
$$= a^{k+1} b^{k+1}.$$

Remember, multiplication of real numbers is commutative.

27. What happens when we try to prove the following: $\frac{n-1}{n+1} < .9$?

This inequality holds when n = 0 (or n = 1 if you want.) So, assume that $\frac{k-1}{k+1} < .9$ and try to show that $\frac{k+1-1}{k+2} < .9$. (Try a few large values of k to see if the statement is even true.)

28. A little bit of matrix multiplication. Prove that $\begin{bmatrix} 0 & 1 \\ 1 & 1 \end{bmatrix}^n = \begin{bmatrix} F_{n-1} & F_n \\ F_n & F_{n+1} \end{bmatrix}$ for $n \geq 1$.

It is easy to see that this identity holds for n = 1, since $F_0 = 0$ and $F_1 = F_2 = 1$. Now, assume $\begin{bmatrix} 0 & 1 \\ 1 & 1 \end{bmatrix}^k = \begin{bmatrix} F_{k-1} & F_k \\ F_k & F_{k+1} \end{bmatrix}$. Then,

$$\begin{bmatrix} 0 & 1 \\ 1 & 1 \end{bmatrix}^{k+1} = \begin{bmatrix} 0 & 1 \\ 1 & 1 \end{bmatrix} \begin{bmatrix} 0 & 1 \\ 1 & 1 \end{bmatrix}^k$$
$$= \begin{bmatrix} 0 & 1 \\ 1 & 1 \end{bmatrix} \begin{bmatrix} F_{k-1} & F_k \\ F_k & F_{k+1} \end{bmatrix} = \begin{bmatrix} F_k & F_{k+1} \\ F_{k+1} & F_{k+2} \end{bmatrix},$$

completing the proof. Look at what can happen! Since $\det \begin{bmatrix} 0 & 1 \\ 1 & 1 \end{bmatrix} = -1$,

$$\det \begin{bmatrix} 0 & 1 \\ 1 & 1 \end{bmatrix}^n = (-1)^n = F_{n-1}F_{n+1} - F_n^2, \text{ or}$$

$$F_n^2 - F_{n-1}F_{n+1} = (-1)^{n+1}$$

[Note: this famous identity is typically proved by induction and is the basis for the equally famous "missing box" puzzle – see Exercise 26.]

This matrix algebra technique allows for an avalanche of Fibonacci identities involving $Q = \begin{bmatrix} 0 & 1 \\ 1 & 1 \end{bmatrix}$:

$$Q^n Q^{n+1} = Q^{2n+1} \text{ gives } F_{2n+1} = F_{n+1}^2 + F_n^2$$

This is done by equating appropriate terms.

29. Here is an example of a proof by <u>strong</u> mathematical induction, where several inductive hypotheses are assumed. The terms of the Fibonacci sequence grow exponentially. We can show that $F_n > \left(\frac{3}{2}\right)^{n-1}$ for $n \geq 6$. We will need to check two separate base cases here. With a calculator, check $F_6 = 8 > \left(\frac{3}{2}\right)^5$ and that $F_7 = 13 > \left(\frac{3}{2}\right)^6$. Also notice that $F_5 = 5 < 5.06$.

Now, assume that $F_k > \left(\frac{3}{2}\right)^{k-1}$ and $F_{k-1} > \left(\frac{3}{2}\right)^{k-2}$. Then,

$$F_{k+1} = F_k + F_{k-1} > \left(\frac{3}{2}\right)^{k-1} + \left(\frac{3}{2}\right)^{k-2}$$

$$= \left(\frac{3}{2}\right)^{k-2} \left(\frac{5}{2}\right) > \frac{9}{4}\left(\frac{3}{2}\right)^{k-2} = \left(\frac{3}{2}\right)^k.$$

The $\frac{9}{4}$ comes from a little bit of backwards thinking.

30. Prove that $T_n = \frac{1}{n+1} + \frac{1}{n+2} + \cdots + \frac{1}{2n} > \frac{13}{24}$ for $n \geq 2$.

This one looks a little peculiar, as the right-hand side is not a function of n. Regardless, we shall proceed by induction and see what happens.

Assume $T_k > \frac{13}{24}$. We will show that $T_{k+1} > T_k$, or that $T_{k+1} - T_k$ is positive. So, let's do the algebra:

$$T_{k+1} - T_k = \left[\frac{1}{k+2} + \frac{1}{k+3} + \cdots + \frac{1}{2k} + \frac{1}{2k+1} + \frac{1}{2k+2}\right] - \left[\frac{1}{k+1} + \cdots + \frac{1}{2k}\right]$$

$$= \frac{1}{2k+1} + \frac{1}{2k+2} - \frac{1}{k+1}$$

$$= \frac{1}{2(k+1)(2k+2)}.$$

The latter fraction is positive for all $k \geq 2$.

31. Let p(n) be the proposition that $\frac{1}{\sqrt{1}} + \frac{1}{\sqrt{2}} + \frac{1}{\sqrt{3}} + \cdots + \frac{1}{\sqrt{n}} > \sqrt{n}$. Let's prove p(n) true by induction for $n \geq 2$.

Assume p(k) to be true. Then,

$$\frac{1}{\sqrt{1}} + \frac{1}{\sqrt{2}} + \cdots + \frac{1}{\sqrt{k}} + \frac{1}{\sqrt{k+1}} > \sqrt{k} + \frac{1}{\sqrt{k+1}}.$$

What remains is to now show that $\sqrt{k} + \frac{1}{\sqrt{k+1}} > \sqrt{k+1}$.

So, follow these steps:

$$\frac{\sqrt{k+1} + \sqrt{k}}{\sqrt{k+1}} > 1 \qquad \text{(easy to see)}$$

$$\frac{1}{\sqrt{k+1}} > \sqrt{k+1} - \sqrt{k} \qquad \text{(multiply by a "conjugate")}$$

--Done--

The following non-induction proof is much simpler:

$$\frac{\sqrt{n}}{\sqrt{1}} + \frac{\sqrt{n}}{\sqrt{2}} + \frac{\sqrt{n}}{\sqrt{3}} + \cdots + \frac{\sqrt{n}}{\sqrt{n}} \geq n = \sqrt{n}\sqrt{n}.$$

Each term on the left is greater than 1, so n of them are greater than n. Finally, divide this by $\sqrt{n}$.

32. Conjecture and prove a formula for $\binom{n}{1}^2 + 2\binom{n}{2}^2 + 3\binom{n}{3}^2 + \cdots + n\binom{n}{n}^2$.

    Collecting data might not reveal a pattern, given the complexity of this sum and the experience we had with a previous, similar problem. So, let's try the reverse and add technique, motivated by the symmetry relation $\binom{n}{k} = \binom{n}{n-k}$.

    $$S = \binom{n}{1}^2 + 2\binom{n}{2}^2 + \cdots + (n-1)\binom{n}{n-1}^2 + n\binom{n}{n}^2$$

    $$S = n\binom{n}{0}^2 + (n-1)\binom{n}{1}^2 + \cdots + 1\binom{n}{n-1}^2$$

    $$2S = n\binom{n}{0}^2 + n\binom{n}{1}^2 + \cdots + n\binom{n}{n-1}^2 + n\binom{n}{n}^2$$

    $$S = \frac{n}{2}\binom{2n}{n} \text{ or } n\binom{2n-1}{n-1}.$$

    Now that we have a nice formula, you could try mathematical induction as an alternate method. See the exercises.

33. Prove by induction that $2^n < \binom{2n}{n}$ for $n \geq 2$.

    Testing this for $n = 2, 3$, and 4 shows that this is not a very "tight" inequality, seen in the results below:

    $$4 < 6, \quad 8 < 20, \quad 16 < 70$$

    In fact, these three inequalities may suggest that a different proof technique might work better. So, here is the beginning of an induction proof:

    Assume $2^k < \binom{2k}{k}$. We will need to show that

    $$2^{k+1} = 2 \cdot 2^k < 2\binom{2k}{k} < \binom{2k+2}{k+1}.$$

    This looks like we will need a lot of factorial manipulation.

128

EXERCISES

Here are a variety of exercises that you can practice on. Always look for alternate, more interesting proofs when you can. Most are designed to be proved by mathematical induction, but some can also be resolved by appealing to the examples provided.

1. $1 \cdot 2 + 2 \cdot 3 + 3 \cdot 4 + \ldots + n(n+1) = \frac{n(n+1)(n+2)}{2}$.

2. $1^2 + 2^2 + \cdots + n^2 = \frac{n(n+1)(2n+1)}{6}$.

3. Conjecture and prove a formula for the sum $F_1 + F_3 + F_5 + \cdots + F_n$ where n is an odd number, and F_n is the usual n-th Fibonacci number.

4. Prove $F_1^2 + F_2^2 + \cdots + F_n^2 = F_n F_{n+1}$. (See picture 10.)

5. Prove that $\log a^n = n \log a$.

6. Prove that 17 is a divisor of $2^n \cdot 3^{2n} - 1$.

7. Prove that $2^{5n} - 5^{2n}$ is a multiple of 7 for $n \geq 0$.

8. Prove that $\binom{3}{3} + \binom{4}{3} + \binom{5}{3} + \cdots + \binom{n}{3} = \binom{n+1}{4}$

9. Prove that $\binom{2n}{n} < 4^n$ for $n = 1, 2, \ldots$

10. Which is bigger: $n!^2$, or $(2n)!$?

11. Which is bigger: $n!$, or $2^{1+2+3+\cdots+(n+1)}$?

12. Prove that $3^{3n+2} + 5 \cdot 2^{3n+1}$ is a multiple of 19 for $n = 0, 1, 2, \ldots$

13. Prove that the sum of the interior angles of any convex n-gon is $180(n-2)$ degrees for $n \geq 3$.

14. Earlier we showed this one using reversing and adding. Now try it by induction: $\binom{n}{1} + 2\binom{n}{2} + 3\binom{n}{3} + \cdots + n\binom{n}{n} = n2^{n-1}$.

15. Prove that $1^3 + 2^3 + 3^3 + \cdots + n^3 = \left[\frac{n(n+1)}{2}\right]^2$.

16. Conjecture and prove a nice formula for $\binom{n}{1}^2 + 2\binom{n}{2}^2 + \cdots + n\binom{n}{n}^2$.

17. Do the same for $1 \cdot 1 \binom{n}{1} + 2 \cdot 2 \binom{n}{2} + 3 \cdot 2^2 \binom{n}{3} + \cdots + n \cdot 2^{n-1} \binom{n}{n}$.

18. Collect data, conjecture a nice formula, and then prove it:
$$\frac{1}{1 \cdot 2} + \frac{1}{2 \cdot 3} + \frac{1}{3 \cdot 4} + \cdots + \frac{1}{n(n+1)}.$$

19. DeMorgan's laws for two sets states that the complement of the union of two sets is the intersection of their complements. State this using good mathematical notation, then state and prove the generalized law for n sets.

20. Prove that $n^3 + 5n$ is a multiple of 6.

21. Determine and prove a formula for the n-th derivative of $f(x) = \frac{1}{x}$.

22. Let $f(x) = xe^{-x}$. Determine and prove a formula for the n-th derivative of $f(x)$.

23. This is a great one for data collection: $F_{n+1}^3 + F_n^3 - F_{n-1}^3 = ?$

24. Prove that $\binom{n}{0} + \frac{1}{2}\binom{n}{1} + \frac{1}{3}\binom{n}{2} + \cdots + \frac{1}{n+1}\binom{n}{n} = \frac{2^{n+1}-1}{n+1}$.

25. Find a nice way to show that $\left(1 + \frac{1}{2}\right)\left(1 + \frac{1}{3}\right)\left(1 + \frac{1}{4}\right) \cdots \left(1 + \frac{1}{n}\right) = \frac{n+1}{2}$.

26. Prove that $F_n^2 - F_{n+1}F_{n-1} = (-1)^{n+1}$. (Try n=6 first.)

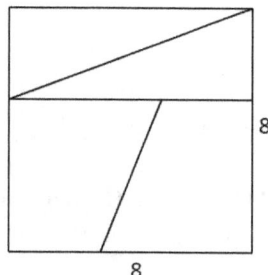 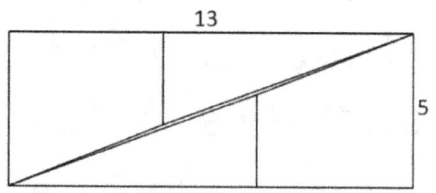

Appendix 1 – Poster Cutouts

The following pages include cutouts to allow creation of your own poster.

___ A. $\binom{n}{2} + \binom{n+1}{2} = n^2$

___ B. $1 + 3 + 5 + 7 + \cdots + 2n - 1 = n^2$

___ C. $1 + 2 + 3 + 4 + \cdots + n = \frac{n(n+1)}{2}$

___ D. $\binom{n}{0}^2 + \binom{n}{1}^2 + \binom{n}{2}^2 + \cdots + \binom{n}{n}^2 = \binom{2n}{n}$

___ E. $F_1^2 + F_2^2 + F_3^2 + \cdots + F_n^2 = F_n F_{n+1}$

___ F. $\frac{1}{2} + \frac{1}{4} + \frac{1}{8} + \frac{1}{16} + \cdots = 1$

___ G. $1 + 2 + 3 + \cdots + n + (n-1) + \cdots + 3 + 2 + 1 = n^2$

___ H. $1 + 3 + 5 + \cdots + (2n-1) = 1 + 2 + 3 + \cdots + n + \cdots + 3 + 2 + 1$

___ I. $1^2 + 2^2 + 3^2 + \cdots + n^2 = \binom{n+1}{3} + \binom{n+2}{3}$

___ J. $(2n-1) + \binom{n}{2} + \binom{n-1}{2} = n^2$

___ K. $1(n) + 3(n-1) + 5(n-2) + \cdots + (2n-1)1 = 1^2 + 2^2 + 3^2 + \cdots + n^2$

___ L. $\binom{m+n}{2} - \binom{m}{2} - \binom{n}{2} = mn$

___ M. $\frac{1}{4} + \frac{1}{4^2} + \frac{1}{4^3} + \frac{1}{4^4} + \cdots = \frac{1}{3}$

___ N. $1 + 3 + 6 + 10 + \cdots + \frac{n(n+1)}{2} = \binom{n+2}{3}$

___ O. $2 + 4 + 6 + 8 + \cdots + 2n = n(n+1)$

1.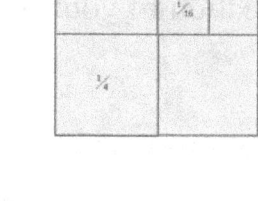

2.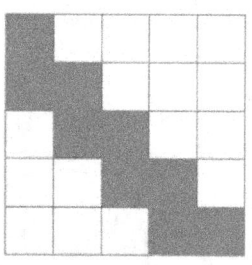

3.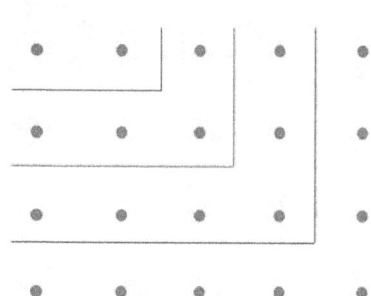

4.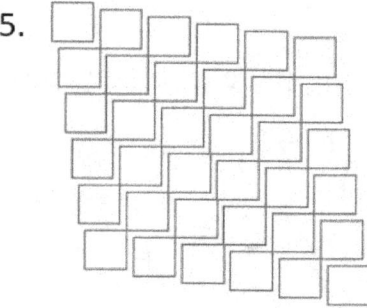

5.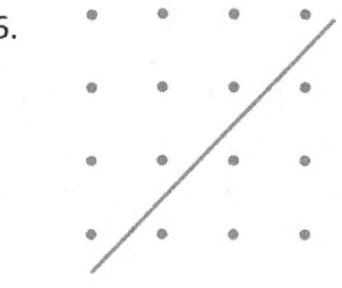

6.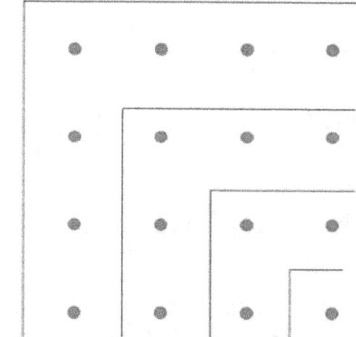

7.

8.

9.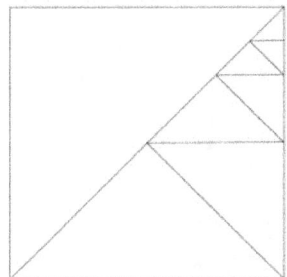

13.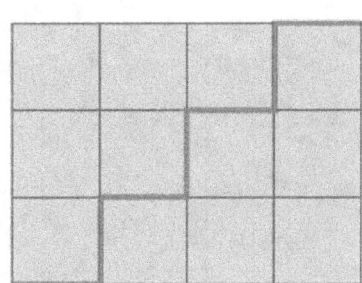

10.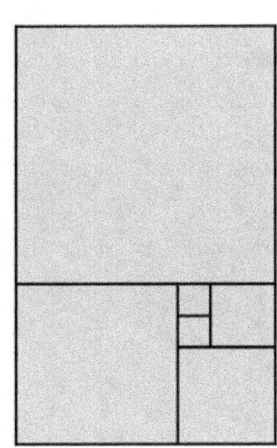

14.
$$\begin{matrix}1&1&1&1\\1&1&1&1\\1&1&1&1\\1&1&1&1\end{matrix} + \begin{matrix}1&1&1\\1&1&1\\1&1&1\\1&1&1\end{matrix} + \begin{matrix}1&1\\1&1\\1&1\end{matrix} + 1 = \begin{matrix}1&1&1&1\\1&2&2&2\\1&2&3&3\\1&2&3&4\end{matrix}$$

15.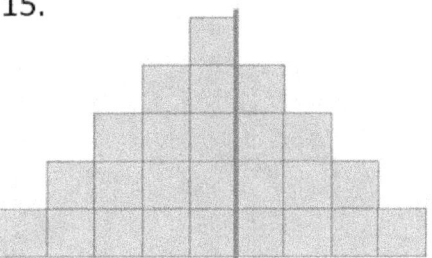

11.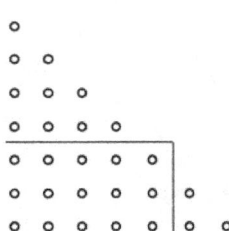

12.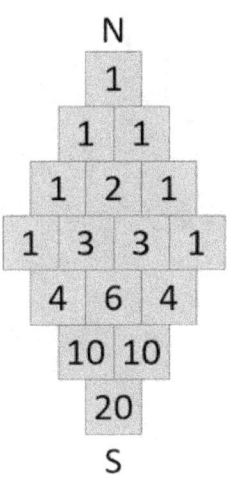

16.

Appendix 2 – Finale

If you liked the 16 picture proofs, here are four late additions to the poster. The following graphics and accompanying formulas were discovered separate from the poster. To keep the previous versions of the poster consistent, we are including them in a separate section.

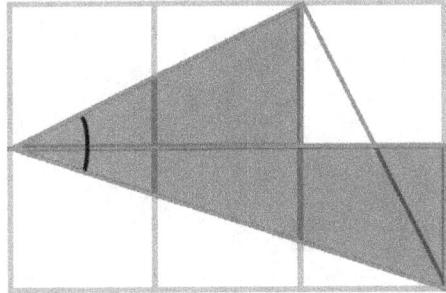

A baby version of the cover.

$$arctan\frac{1}{2} + arctan\frac{1}{3} = \frac{\pi}{4}$$

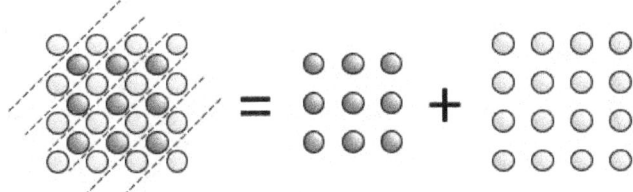

$$1 + 3 + 5 + 7 + 5 + 3 + 1 = 3^2 + 4^2$$

$$\vdots$$

$$1 + 3 + \ldots + (2n-1) + (2n+1) + (2n-1) + \ldots + 3 + 1 = n^2 + (n+1)^2$$

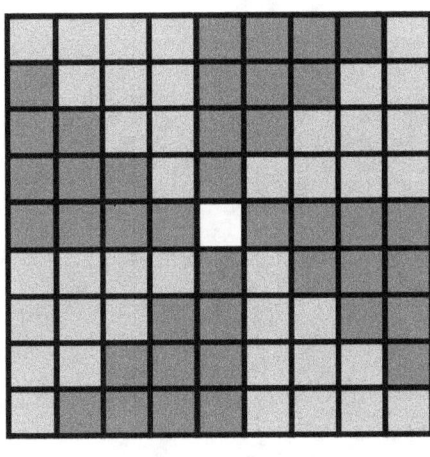

$$(2n+1)^2 = 8T_n + 1$$

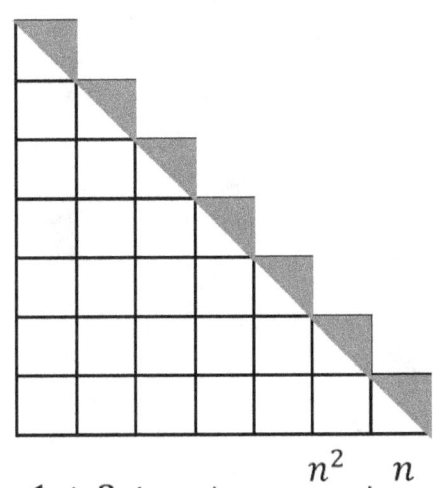

$$1 + 2 + \cdots + n = \frac{n^2}{2} + \frac{n}{2}$$

About the Authors

"My father's statistics textbook, written in the late 1970's. The cover picture served as inspiration for writing in general, and this textbook in particular."   -- Dean Zeller

-- RMG

Dr. Richard Grassl

Richard has degrees from Santa Clara University, University of Oregon, and the University of New Mexico. He was a faculty member at UNM for 20 years. Afterwards he taught at Muhlenberg College in Pennsylvania, holding the Truman Koehler Professorship for two years. Grassl taught at the University of Northern Colorado for another 20 years, during which he served as chair of Mathematical Sciences for 14 years, and briefly as an assistant dean of his college.

After retirement in 2011, he continued working, first as a visiting emeritus faculty at Cal Poly SLO for five years and then back to UNC. He is now in his 5th year, teaching mostly MA courses for in-service teachers on ZOOM.

He directed statewide mathematics contests for 7th–12th grade students for over 40 years at UNM and UNC, published 37 research articles in mathematics and mathematics education, several textbooks, was active in scholarly presentations, and was the Recipient of the 2009 Rocky Mountain MAA section Teaching Award.

He has worked with various Mathematics Teacher Circles and is currently a member of the MTC Network Advisory Board.

Dean Zeller

Dean Zeller started college in 1987 at Bowling Green State University in Ohio, as a theater major, with goals and aspirations of public performance for the rest of his life. Young Zeller discovered he was an awesome actor, except for the fact he couldn't sing, dance, act, or memorize lines. It was then that he switched to his second career direction, mathematics and computer science. It has been a rollercoaster of a trip since.

While his goals and aspirations changed, the desire to perform was still strong, despite lacking the skills necessary to succeed in theater. To continue his love of performance, he became a teacher instead, following in the footsteps of both parents. Teaching students instilled the same sense of professional satisfaction as performance on a stage, and the pay was much much better.

Since then, Zeller has had a nearly 25-year career in teaching, serving as faculty at Bowling Green State University, Park University in Missouri, the Art Institute of Jacksonville, and the University of Northern Colorado. Zeller has researched advanced methods of algorithm computation, including numerical analysis, data mining, and evolutionary computation. He has also explored the now booming fields of artificial intelligence and graph theory. He remains dedicated to the further development of computer science at the college and K-12 level.

Dean runs the firespinning performance group, Fahrenheit 360°. They have performed at FridayFest in Greeley, Colorado, for the last six years. Dean also enjoys hiking, wilderness survival, culinary arts, comic books, and learning new skills. He would like to further his expertise with drones, if given the chance.

www.ingramcontent.com/pod-product-compliance
Lightning Source LLC
Chambersburg PA
CBHW080458220526
45465CB00006B/2308